(handwritten inscription)

To Herman Winick for the
champion of science for the first 80 y[ear]
in occasion of his first 80 y[ears]
With esteem & friendship
Claudio
Annau, 7 November 2012

Claudio Tuniz is a world-renowned expert in geochronology using particle accelerators. He is Assistant Director of UNESCO's International Centre for Theoretical Physics in Trieste, Italy, where he promotes the use of atomic and nuclear physics in palaeoanthropology. He was director of the accelerator dating centre at the Australian Nuclear Science and Technology Organisation and has published widely on Australian prehistory.

Richard Gillespie built a radiocarbon laboratory at the University of Sydney before taking up research positions at Oxford University, the University of Arizona and the Australian National University. He is an authority on dating bones and shells, with wide fieldwork experience in Africa, North America and Australia.

Cheryl Jones is a science journalist who for many years has covered developments in Australian prehistory for international and Australian media, including *The Australian Financial Review*, *The Canberra Times* and *The Bulletin*.

The BONE READERS

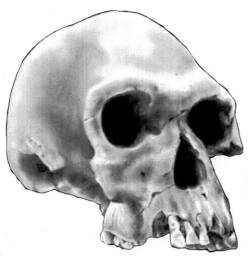

Science and
Politics in
Human Origins
Research

Claudio Tuniz
Richard Gillespie
Cheryl Jones

Left
Coast
Press
inc.

Walnut Creek, California

LEFT COAST PRESS, INC.
1630 North Main Street, #400
Walnut Creek, CA 94596.
http://www.LCoastPress.com

Library of Congress Cataloguing-in-Publication Data available from the publisher.
ISBN: 1-59874-475-5
        978-1-59874-475-0 paperback

Illustrations by Mario Tiberio and Walter Gregoric
Index by Garry Cousins
Set in 11/15 pt Sabon by Midland Typesetters, Australia
Printed in the United States of America

♾™The paper used in this publication meets the minimum requirements of American
National Standard for Information Sciences—Permanence of Paper for Printed Library
Materials, ANSI/NISO Z39.48–1992.

09 10 11 12 13   5 4 3 2 1

For Maude, Patrizia
and the late Diana

Contents

Junette

Junette Mitchell did not hesitate when asked why she had given a DNA sample to a geneticist studying the evolutionary history of Australian Aborigines. 'I wanted to see how close we were to Mungo Lady,' said the quietly spoken elder of the riverine Paakantji people from south-western New South Wales.

Her motives when she gave the sample were strong enough to overcome the suspicion many Aborigines have of genetics research. They were also a match for opposition to 'colonial science'.

Mitchell's people, who have a history of frontier conflict and dispossession stretching back more than 150 years to the time when Europeans were encroaching on their territory around the mighty Murray, Darling and Lachlan rivers, are among traditional owners of the Willandra Lakes Region World Heritage Area, about 800 kilometres west of Sydney. Her country takes in the relics of a 1,000-kilometre-square system of five huge lakes, dry for 18,000 years and now covered with saltbush and mallee scrub. The Paakantji now lead tour groups to the Walls of China, a 30-kilometre-long lunette, or crescent-shaped sand dune, which rises up to 40 metres from the eastern and southern shores of Lake Mungo, the centre of the system. They also work as land and heritage managers at Mungo National Park, which attracts 50,000 tourists a year.

Dotted with strange forms beaten by the westerlies into white quartz sand, the vast lunettes around the lakes have delivered up the skeletons of more than 100 ancients. Mungo Lady illuminates an ancient culture and its interaction with a new land. She was strolling around Lake Mungo when the first modern humans were venturing

1

into Europe. Her bones, along with those of her contemporary, Mungo Man—claimed in 2001 to have yielded DNA—are the oldest on the continent. She is the world's oldest known cremation, and she lies at the heart of arguments about the date of the first colonisation of Australia. She is also at the centre of the wider debate on human evolution—whether our species evolved recently from an 'African Eve' or had a more ancient and complex origin. According to the ascendant 'out of Africa' model, our species evolved in Africa up to 200,000 years ago and spread out across the globe, replacing the descendants of even earlier African migrants. The rival 'multiregionalist' model has *Homo sapiens* evolving at several points on the globe, with interbreeding pushing our species down the same evolutionary pathway. Just how the first Australians, Europe's Neanderthals and Indonesia's 'hobbits' fit into the global human evolutionary scheme is critical to the argument. Australia's multiregionalists engage in heated debate with its Africanists, and sometimes the arguments take on ideological aspects. According to archaeologist Hilary du Cros, 'Indigenous communities in Australia will probably be barracking for the multi-regional model as their creation myths tell them that they have been always here.'

The burials, hearths, shell middens and unique geomorphic features that won the Willandra UNESCO World Heritage status in 1981 have, at times, been a battleground for scientists and the region's Aboriginal communities—the Paakantji, Mutthi Mutthi and Ngyiampaa—which once formed an alliance called the 3TTGs (Three Traditional Tribal Groups). Some local Aborigines have vehemently opposed research, and some had rejected the geneticist's request for DNA samples. 'There's a lot knocked her back,' says Mitchell. Asked to comment on opposition to genetics research, she said: 'It's only up to you—if you want to find out how close you are to those remains. That's what we wanted to do.'

Mitchell works tirelessly to pass Paakantji 'lingo' down to her people's children. One method is a game called 'Paakantji whispers'— the children sit in a circle with Mitchell and pass on new words. She is scandalised by some of the older children who have picked up naughty words in lingo and are repeating them at school, all too often when giving cheek to uncomprehending teachers. She also wants to record a Paakantji story in an illustrated book for the kids, but she would not give much away when asked about her creation beliefs. Her views on her people's origins differ markedly from those of the geologists, dating experts, geneticists, biologists and archaeologists who have been making pilgrimages to the Willandra since 1968, when geologist Jim Bowler discovered Mungo Lady's remains eroding from dunes on the Joulni sheep station on the southern shores of Lake Mungo. At Mungo, the scientific world view stands alongside the traditional one. It's an arrangement that sometimes works, but often the politics of the past intervenes.

Where did the Aborigines come from? 'Well, they never floated over,' Mitchell says firmly. 'A lot of people always say they floated over. But, eh! I always think their heart was little people. Monkeys and apes and gorillas and that—we never came from them—but we come from these other little people.'

Here? 'In Australia. I believe it from my mother, from her grandmother—handed down, you know? We couldn't float over here, because the sea was too rough. But we was here all the time ... That's as far as I'll go.'

Still, Mitchell was curious to see what the genetics said of her kinship with the mysterious Ice Age lady who harvested mussels and fish from the lake tens of thousands of years ago. Already, scientific research on Mungo Lady had confirmed her belief that Aborigines had been in Australia 'a long, long time'.

'It's a big breakthrough now to prove that Aboriginal people was in Australia before anyone had 'em here,' she says. 'That's what Mungo Lady showed ... They said she was Paakantji but it's hard to tell. So I was thinking I'd like to find out properly.'

Mitchell made the comments in an interview in the yarns tent at the Mungo Festival, held in 2006 to celebrate the twenty-fifth anniversary of the Willandra Lakes Region World Heritage listing. In a later interview, when asked about another prehistory political hotspot— what drove the Australian megafauna, like the bird *Genyornis*, and the two-tonne marsupial, *Diprotodon*, to extinction—she said: 'I don't think [it was] the Aboriginal people. You know, they used to kill. They

would have a big feast. That was mostly on the smaller animals—
fish, turtles, yabbies, mussels. When we go up the river, we go for the
smaller animals; we don't go and kill out the big buck kangaroo. We
only go for the young stuff.' And the Aboriginal practice of burning
the landscape was not responsible for extinctions either; rather, it was
critical to the regeneration of the bush.

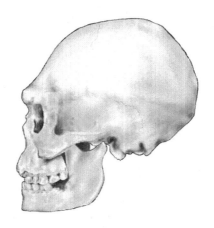

I
LANDFALL

1 Timelords and god-scientists

Tourists skated as much as drove along 100 kilometres of wet, red clay from Mildura to the national park to attend the 2006 Mungo Festival after a rare night of heavy rain in a place that averages just 250 millimetres a year. The bitumen roads around the Riverina end abruptly with the irrigation channels that sap the ailing Murray, its murky waters choked with weed and feral fish, to coax out of the semi-arid zone crops from another hemisphere. Fields of bright yellow canola flowers clash with dull green remnant eucalypts. Then the incongruous landscape of vineyards, citrus orchards, cotton fields and pampered dairy cows gives way to impossibly flat, sparse woodlands of bonsaied trees and native tussocks. Eventually, the road dips to the surface of Lake Outer Arumpo, the southernmost larger lake of the network, before rising over that lake's lunette and on to Mungo.

In the drizzle outside the Mungo Visitors' Information Centre, a local Aborigine gave a drum of eucalyptus leaves a liberal dose of kero, to enhance their natural flammability, before setting them alight for the smoking ceremony. Bathing yourself in the smoke would keep you safe, the crowd was told. Local didgeridoo maestros, their instruments' natural tones ratcheted up by powerful amps, were felt as much as heard. Traditional dancers wearing only loincloths froze as they performed in the howling wind before busloads of tourists.

'It all comes about because of that Mungo Lady—our ancestor, our mother, who came up out of the ground, who was accidentally found, who was taken away ... went around the world, mind you, so that white man could study her,' Mutthi Mutthi elder Mary Pappin said in her welcoming address. 'That continuation of our culture defies all kinds of beliefs throughout the world. The Australian Aborigine was very clever to survive in a harsh environment and continues on. We know that our ancestors today are standing with us all and encouraging us to make sure that we will continue our cultural heritage.'

Some of those white men were in the crowd. They had been at the 'Legacy of an Ice Age' conference, held in the national park in the lead-up to the festival. (Scientists don't always see eye to eye with Pappin but like to tell anecdotes about the feisty, diminutive lady who once drove would-be sand miners out of town, accusing them of wanting to rape Mother Earth.)

'An important feature of this meeting is engagement between scientists and the three Traditional Tribal Groups,' the conference pre-publicity had said. 'Group forums will seek the views of the traditional custodians, explore the interaction of indigenous people and science, and discuss the management of this heritage.' Scientists and archaeologists, some of them veterans of the early days of research at the site, answered the call. They shivered alongside representatives of the 3TTGs in a marquee as presenters, competing with the coughs and splutters of the generator outside, gave papers, many of them recounting or synthesising earlier research, with new studies limited in the political climate of the past two decades.

The Mungo burials were swept up in the wave of protest over Aboriginal remains collected in the nineteenth and early twentieth centuries and still held in museums and universities around the world. Indigenous people, who had had bad experiences with the amateur pseudo-scientists of the first century of colonisation, were, from the 1970s, undergoing a 'decolonisation of the mind'. They were demanding ownership of the past—control over what research was done and a greater say in how it was interpreted. They were working to preserve or resurrect traditional customs and beliefs, some of which conflicted with the scientific view. The battle to get indigenous remains and artefacts repatriated to Aboriginal communities began, with the most hardline stance in Tasmania, where first contact with Europeans had been bloody. Indigenous world views today vary widely, ranging from traditional through Christian and New Age to scientific, or a combination of them all. Many Aborigines agree to disagree with the scientists, and their stance on research, at least in the communities, ranges from opposition to ambivalence and support. The 3TTGs have blocked research on Mungo Child, probably a contemporary of Mungo Man and Mungo Lady, and discovered at Joulni in the late 1980s. The find coincided with protest ignited by celebrations in 1988 marking the bicentenary of the British colonisation of Australia. The bones have remained in the dune, first covered with a sheet of corrugated iron and later protected with shade cloth and sand. There was talk of a salvage excavation, but it came to nothing.

Still, there were signs of a thawing in relations in 2006.

Delegates heard reports on newly discovered fossil footprints left 21,000 years ago. And the masculinity of Mungo Man, or Willandra Lakes Hominid 3—WLH3—or LM3, as he is variously known in the scientific literature, was restored, according to one researcher, following a review of his vital statistics. Palaeoanthropologist Alan Thorne recounted how almost 40 years before he had spent six months gluing the fragments of Mungo Lady's skull together. And the megafauna extinction debate was rekindled. Did overhunting, firing of the landscape or climate change drive the big animals into oblivion?

Outside, an Aboriginal elder, a joey at his feet, basking in the sun, glanced at the marquee and observed dryly: 'They're better off in there than making bombs.'

Researchers and elders were loaded into mini-buses and taken to key sites around the lakes on an excursion setting the scene for yet another cultural clash—between science and art. Inevitable art installations had been deployed along the Mungo tourist drive for the festival. The works, exploring themes ranging from time to the 'spirit of place', left many scientists scratching their heads. One work, which had blown over and broken, was blocking the path to the Walls of China. Perhaps the destruction was intentional, symbolising the problem of 'preservation' of bones and archaeological material in the ever-changing environment of Willandra. Perhaps it was a statement that, in life, there's always something in the way.

A visit to the Joulni sites took on aspects of Anzac Day. Conference delegates gathered on the lunette overlooking the site of a blowout, a place where wind and rain had exposed older strata more than 30 years before. A star picket in front of a 'residual'—a sculpted remnant dune—marked the site of the Mungo Lady burial. Another 450 metres to the east designated the resting place of Mungo Man, found by Bowler in 1974. This was Aboriginal land. The 3TTGs held the lease over the old sheep station, and the public was banned from the area. It had significance, too, for the researchers who, in the early days, dug the sites in an area described as 'Australia's Rift Valley', a zone which to Aborigines is the centre of creation, and which to scientists holds the key to understanding the evolution of our species.

There was one, a former cattle musterer from the Snowy Mountains, who is now one of the giants of geology. Jim Bowler translated the story written in the sediments, a story he says changed his life. It is encoded in coarse gravel dumped by big waves whipped up by the westerlies, in fine quartz beach sand hurled onto the lunette by the wind, in tiny grey clay pellets dislodged from the lake floor in dry times, in wüstenquartz—red desert dust—swept onto the dunes as the arid Centre expanded, and in soils that form when dune building ceases. In the early days, the radiocarbon dating method revealed that the burials were very old. A second dating revolution, based on the liberation of energy from grains of sand, later pushed back the age. The synthesis of the data told a familiar story. It was about Australians struggling with aridity.

Archaeology students arrived on the lunette. Only a few were Aboriginal, but enough to swell the ranks of indigenous archaeologists, at the time of the Legacy of an Ice Age conference numbering only about 10. Then the father of Australian archaeology said his piece.

John Mulvaney founded an Australian prehistory department at the Australian National University in the 1960s and wrote the first textbook on the subject. An outspoken conservationist who fought hard battles to protect Aboriginal heritage in Tasmania and Kakadu National Park, he had campaigned to get the Willandra Lakes region its World Heritage status. In 1965, he populated the Pleistocene with people, with the discovery of the first site dating from that epoch— Kenniff Cave on Mount Moffatt Station in Queensland, where a dated sequence of stone tools extended nearly 3 metres deep to about 22,000 years ago. Later, along with Wilfred Shawcross, he dug at Mungo. It was to be his last archaeological excavation.

On his return to Mungo in 2006, by then in his eighties, he could not resist a gentle dig at the three generations of dating experts assembled— radiocarbon specialist Richard Gillespie, and Rainer Grün, John Prescott and Matt Cupper who use newer methods. Many archaeologists worry about losing prehistory completely to these practitioners of the arcane sciences—the 'timelords'.

'I do want to say, particularly to the scientists who think only in dating, what is the significance of this Mungo site,' he said. 'It dates early occupation, but it does far more. These burials here—reliably dated now—we have a cremation, we have an inhumation [burial] in which the corpse has been sprinkled with considerable quantities of ochre. The ochre, as far as we know, had to come [from] at least a couple of hundred kilometres from here. Forty-two thousand years ago, already people knew enough about the country, knew enough about geology ... to know about this ochre. The point about the burials is that they are human actions. That is to me the real significance of Mungo. In the nineteenth century, indigenous people were thought to be just savages—they were incapable of counting, incapable of draw- ing—they were sub-human. Here we have, at 42,000 years ago, people who were burying the dead. We don't know for what reason but they seem to be human values—respect and love of the dead, fear of the dead and a particular process—you burn the body and you smash them up and bury them in a hole. What's the significance of that? Why do that? Or why cover the corpse with ochre? People obviously have thoughts about the afterlife.'

**

When Mulvaney began his work, the study of the human past was still based in humanities faculties of universities world wide. Then a 'new archaeology' or 'processual archaeology' swept academia. New archaeology propelled the field towards the natural and social sciences and viewed humans as part of the ecology. Archaeologists were adopting Enlightenment philosophy—positivism, empiricism, rationalism and reductionism—the epistemology that grounds science. They were coming up with theories to explain change in the artefacts and bones in the archaeological record, and had a quiet confidence that the scientific method was the most reliable path to knowledge—a way to test their hypotheses. Mulvaney was ambivalent about the new philosophy influencing the second generation of Australian prehistorians. 'I remain a humanist,' he said in 2008. 'No matter how many scientific tests you do on bones, they won't reveal anything about human creativity. You can't treat humans as animals.'[1]

The shift to new archaeology came in the early stages of the radiocarbon revolution, at a time when the dating experts—the timelords—were just beginning to illuminate the past more brilliantly. Other quantitative methods were being developed, too.

Then another philosophy, the ideology behind it rooted not in the Australian desert or in laboratories but in Paris, would influence the field. A new group of prehistorians, some of them qualified in science, drew on ideas taking shape in the politicised, postmodern air of university arts faculties. Borrowing from French literary theory and supporting the 'post-processual' theory pioneered by the British archaeologist Ian Hodder, postmodernists deem truth to be a myth, research to be irredeemably biased, science to be 'just another text'—a Western construct—the past to be unknowable, and politics to be more important than knowledge. Many deride, or 'deconstruct' as 'racist' or 'sexist', the scientists or archaeologists with whom they disagree.

It is impossible to gauge the impact of postmodernism on prehistory, and many scholars and university departments have let the wave wash over them. However, there have been accents of the philosophy in cultural heritage management and some academic debates.

Tim Flannery, a scientist accustomed to calling a spade a spade and well known for his claim that a single hunting blitzkrieg wiped out the Australian megafauna, has been viciously attacked because of the supposed risk that his views could be used by anti-Green forces. Mulvaney, meanwhile, has been attacked for opposing the reburial of ancient remains. And an entire Honours project was devoted to 'deconstructing' the three editions of his textbook, *Prehistory of Australia*, the last co-authored with archaeologist Johan Kamminga.

In her 2005 thesis, 'Specimens and Stone Tools: Aboriginalism and depictions of Indigenous Australians in archaeological textbooks', Flinders University's Belinda Liebelt claimed to have unmasked meanings in the texts that 'subjugate and oppress Indigenous people's knowledges of themselves and their pasts, as inadequate and unscientific'.

Most researchers with a scientific world view agree that their discipline is on some level influenced by social and cultural factors. 'That is not the same as saying the knowledge which is being generated is the same as that in the horoscope in the back of the *Women's Weekly*,' says La Trobe University's Tim Murray.

Complicating the issue are rivalries between research teams, universities and museums as they compete for glory, jobs and dwindling funds. Claims for sites to which researchers have devoted years of work are guarded jealously against counterclaims from rival research teams presenting new data. One of us, Jones, has written elsewhere that 'archaeology without science is so much poetry'. Prehistory relies on science for dates, and for reading the complex histories written in sediments, fossils and DNA. Big research teams typically include scientists as well as archaeologists. But when a scientific paper is published, some players adopt the tactics of the political activist, flouting the self-correcting protocol of science involving peer review and replication of results. They line up on factional grounds. Debate rages in the media and public fora, in preference to learned journals. University and museum public relations machines are cranked up. Factionalism and lack of rigour have led to some internationally renowned blunders.

The big debates in Australia are echoed in the United States, Canada and the Pacific islands. The North American continent shares Australia's curious ecological history, with its larger animals becoming extinct soon after people arrived in force. In the United States and Australia, despite what science has to say about these extinctions, sometimes politics and ideology obscure the scientific evidence. And the repatriation debate reverberates through countries where first peoples have a history of dispossession and disadvantage. In the US, it centres on the bitter dispute over the 9,000-year-old Kennewick Man remains. Discovered along the banks of the Columbia River in Washington State in 1996, Kennewick Man is said to resemble populations such as the Ainu, the indigenous people of Japan, more closely than modern Native Americans. The high-profile dispute has played out in courts, parliaments, academic journals and the media.

This book is not a treatise on formal epistemology, and we spare the reader the obscurantism of Foucault and Derrida. We will, however, recount the collision between researchers and politics in the quest for answers to four big questions that have reverberated through Australian prehistory studies: Where did the Aborigines come from? When and how did they get here? Who were they? How did they interact with the environment?

Some of the results and hypotheses reported here will be refined—perhaps even abandoned—as more evidence comes to light. Our focus will be on Australian prehistory but we will show how Australia is critical to global questions about our species' deep past, including the evolution and dispersal of our genus during the Quaternary, the most recent geological period. The book involved years of research, visits to sites and laboratories around the world, attendance at conferences and interviews with some of the world's leading experts.

***

By the end of 2006, hopes were high for a resurgence of research in the Willandra. Scientists, led by ANU dating expert Rainer Grün, and the 3TTGs had won $735,000 in government grants for a grand three-dimensional survey of the region aimed at improving conservation of the skeletons, hearths and middens eroding from the dunes. The surveyors would deploy satellites, lasers, aircraft, deep physics, geomorphology and traditional knowledge. The push for the project followed the destruction by wind and rain of a skeleton that had been exposed in the sand. The scene of destruction had upset elders, but research on human remains was still off the agenda. And apart from salvage excavations in urgent cases to be performed in a separate project led by the 3TTGs, only in situ conservation was sanctioned by traditional owners. Still, there had been hopes that a 'keeping place' to house remains at the Willandra would smooth the path to research.

However, there have been alarming signs that politics is weakening one of the foundations of science—peer review, a system of checks and balances central to the scientific method. The British journal *Nature* and the American journal *Science*, which compete for top ranking in the scientific food chain, along with hundreds of other refereed journals, subject manuscripts to peer review before publishing the research. In theory, independent scientists eminent in the relevant field vet the research on the basis of its scientific merit. The journals accept or reject papers on the advice of referees and the response of the scientists, from whom a referee's identity is usually kept secret. Science funding agencies put research grant applications through the same gruelling process when deciding whether to back them.

One referee described a grant proposal made by a team of elite researchers as 'the fly-by-night neo-colonialism of god-scientists' putting samples through 'their expensive machines'. Another grant proposal—for a project critically dependent on the interpretation of the subtleties of deep physics—was attacked by a reviewer because no humanities-trained archaeologists were included on the research team. '... [T]his is a very arrogant application,' the reviewer wrote. '... [T]his is an application by scientists treating archaeologists with contempt,' continued the review, which implied that the applicants—another team of elites—were mere technicians.

The disadvantage of indigenous people, their rights and the solutions to their problems are political matters. What happened in the deep past, however, is a scientific question. The idea that modern ideology determines reality 50,000 years ago is a hypothesis unworthy of refutation. Like their compatriots, Australian scientists and archaeologists agonise over the nation's race relations. Some have been attacked for 'playing into the hands' of racists by releasing results seen as politically sensitive, but they know that there is nothing repugnant in the prehistory emerging through their work.

This is the story of physicists, chemists, geologists, dating experts, palaeontologists, palaeoanthropologists, geneticists, biologists, palaeo-ecologists and archaeologists illuminating the deep past. They use the scientific method. It is a way of looking at the world through experimentation and observation in an approach pioneered by the likes of Galileo Galilei and Francis Bacon 400 years ago. They subject their data, analysis and interpretation to the rigorous scrutiny of their peers, and often end up in heated debates among themselves. Some of them use isotopes generated from the radiation from exploding stars to date the arrival of Australia's first people. Others use the eggshell of extinct birds and ancient pollen in deep sediments to assess the Aborigines' environmental impact. Still others read the record of human migration from the blood pulsing in our veins. They are the 'timelords' and 'neo-colonial god-scientists', and this book is about what they can tell us.

2 Heat and light

'Walk towards my voice.' Luminescence dating expert Richard Roberts is 'showing off' his laboratory. It is pitch-black.

A colleague is cleaning up some samples in strong hydrochloric acid, working under a point of dim red light. The 'sensory deprivation lab' disorients all but the initiated. Here, Roberts dates ancient bones and artefacts by hitting with heat or light the grains of sand in which the objects were buried. The samples, carefully collected without exposure to light, must be kept in the dark before the experiment.

The quartz crystals have absorbed energy from radioactive elements in the earth—potassium, uranium and thorium—and from cosmic rays. The energy has knocked some of the sand grains' electrons out of their usual positions, and they have accumulated in defects in the crystal lattice in numbers proportional to time. The electron traps are often sites where impostor atoms have ousted atoms of silicon, installing themselves in their place and distorting the distribution of charge around the lattice.

In the next room, also pitch-black, a sophisticated instrument focuses green laser light on a batch of sand grains, forcing the wayward electrons back to their normal positions. They surrender their excess energy in the form of ultraviolet light, the intensity of which betrays the time since the grains were last exposed to sunlight before being buried. Optically stimulated luminescence (OSL) dating is a variation on thermoluminescence, a method using heat to force the electrons back to their ground states.

The discovery of luminescence dates from 1663, when British chemist Robert Boyle took a diamond to bed. He observed that the stone glowed on contact with 'a warm part' of his body, in what must have been one of the most fun experiments of the seventeenth-century scientific revolution. In 1885, German physicist Eilhard Wiedemann noticed that the irradiation of various crystals with the mysterious cathode rays, now known to be electrons, sparked luminescence. Experiments on 'cold light' paved the way to the discovery of x-rays by Wilhelm Conrad Röntgen in the same year, and of natural radioactivity a year later. In 1953, the American chemist Farrington Daniels and his collaborators proposed the application of thermoluminescence to archaeological dating. Martin Aitken of Oxford University used the method in the early 1960s to date ceramic materials found at archaeological sites. Technological advances in photonics over the past 50 years have increased the precision and range of luminescence dating. The method has a reach of half a million years, well beyond the 'radiocarbon barrier' at 50,000 years, so Australia, with major turning points in archaeology beyond that limit, has been quick to take up the method.

But while the lab work is grounded in the dispassionate hard sciences, it has thrown Roberts, of Wollongong University, and other scientists, into political minefields.

People first arrived in Australia late in the Quaternary, which started 2.6 million years ago. The Quaternary spans the Pleistocene and Holocene epochs, and is distinguished by the emergence and spread of our genus, *Homo*, and the start of the ice age cycles that continue today. And it is a time of much action on the geological timescale. Hard-rock geologists colonising deep mineral-rich strata disparagingly call it 'the dirt on top'. Some of them, at the International Commission on Stratigraphy (ICS), have attempted to wipe the Quaternary from the geological timescale, subsuming it into the Neogene. The move has incensed geologists in the International Union for Quaternary Research (INQUA), who say it is designed merely to make the geological timescale tidier. (The ICS is charged with hammering 'golden spikes' into the boundary of each stage of the geological record.) Australians, including Brad Pillans, a palaeomagnetic dating expert at the ANU, have been in INQUA's front line of defence. Meanwhile, academic

feminists congregating in the warmer conditions of the Holocene—
recorded in the very top sediments spanning the past 12,000 years—
have had a go at the Pleistocene. They have condemned it as the
preserve of males bashing around the outback to answer 'big questions' carrying the most international kudos—like the date of human
colonisation of Australia.

'The majority of work done on the Pleistocene is done by men and
... Holocene research is popularly believed to be the "soft option" in
Australian archaeology,' wrote scholars Laurajane Smith and Hilary
du Cros, in a volume of papers presented at the second Australian
Women in Archaeology conference held in Armidale in 1993.

> The last twenty to thirty years could certainly be described as the
> period when archaeologists (mostly male) were constantly excavat-
> ing (mostly Pleistocene) sites in rural and outback areas. It is impor-
> tant to note that in the 1960s and 1970s archaeological research was
> seen to be done mostly by male researchers, and that these researchers
> were interested in the 'big questions' that were seen to have worldwide
> significance. They were also researching in regions or areas which were
> seen to be privileged either because of their extreme age (i.e. Pleisto-
> cene) and/or because they were in rugged arid country—country that is
> strongly associated in Australian popular culture with machismo and
> raw courage—and is not seen as the place for women. Further, major
> breakthroughs for researchers during this period have largely been
> associated with radiocarbon dating, and this has dominated the type of
> archaeology conducted in Australia. One of the most significant events
> in Australian archaeology was the dating of a hearth at Lake Mungo
> to about 32,000 BP [36,000 calibrated]. The finding and dating of
> Pleistocene sites is still seen today as a major archaeological event, with
> new researchers being welcomed into the 'Pleistocene club' following
> their first confirmed Pleistocene date.

ANU archaeologist Rhys Jones was a member of the club, along
with the growing body of dating experts. One of us, Claudio Tuniz,
another self-confessed member, who led one of the two Australian
accelerator mass spectrometry (AMS) radiocarbon centres in the
1990s, recalls the adrenaline rush when obtaining a date of 36,000
years for Ngarrabullgan, a Cape York site studied with archaeologist

Bruno David, now at Monash University. When Tuniz joined a team, led by Roberts and including Rhys Jones and Mike Morwood, dating the controversial Bradshaw rock art of the Kimberley, he was warned by ANU radiocarbon dating specialist John Head about the slippery road of Australian prehistory politics. Roberts was using the OSL technique to get a minimum age for the art by dating sand grains in mud wasp nests overlying it.

Roberts first used the luminescence method when he was investigating the impact of tailings from the Northern Territory's Ranger uranium mine on one of Australia's environmental treasures—Kakadu National Park. He was dating sand aprons in the Magela Creek catchment by thermoluminescence. He contacted Jones, proposing to pool results on archaeological sites on the formations, and to do further dating on the lowest artefacts scattered in the deposits at levels with no charcoal suitable for radiocarbon analysis.

Rhys Jones, a Welsh scholar educated at the University of Cambridge, was a Renaissance man and a raconteur with a singular sense of humour. Shortly before his death from leukaemia in 2001, he told how a journalist had approached him for an interview so that he could write an obituary in advance. Jones granted the request. He had arrived in Australia in 1963 in the golden age of Australian archaeology. He later suspected that a radiocarbon barrier, at about 40,000 years, was depressing dates on Australia's oldest sites. If he was right, the ancestors of the Australian Aborigines might have been the first modern humans out of Africa, and Australia would graduate from a 'backwater in the global debate on human origins' to the centre of the argument. Jones despaired that his discipline was surrendering to the scientists and losing its vital links to the humanities. He voiced his views on the tension between science and archaeology in an archaeometry (archaeological science) symposium at the Australian Museum in Sydney in 1982. In a paper, 'Ions and eons: Some thoughts on archaeological science and scientific archaeology', he concluded that 'if archaeometry is not archaeology, it is nothing'.

But, trained in the natural sciences as well as the humanities, he also feared that postmodernism was weakening archaeology. He was booed at a conference at his alma mater, Cambridge University, when he hypothesised about a museum curator under pressure to present traditional world views as fact—that Aborigines had always been in Australia; they were autochthonous; they sprang from the land.

Sometimes science supports traditional beliefs. Some northern Australian traditions, in which Dreaming figures arrived on the continent from across the sea, tally with the scientific view of colonisation from South-east Asia.[1] '... [A]rchaeological knowledge is underpinned by the discourse of logical positivism that stresses objectivity and rationality. This discourse ... underwrites the authority of archaeological pronouncements while, at the same time, devaluing the authority of Indigenous knowledge,' writes Laurajane Smith, who is a specialist in cultural heritage management based at the University of York.

Jones jumped at Roberts's offer to deploy luminescence dating, and the pair teamed up with archaeologist Mike Smith, now of the National Museum of Australia, a sceptic who had been working one of the oldest desert sites, the Puritjarra rock shelter near the MacDonnell Ranges in central Australia. Jones moved easily between Aboriginal and European cultures. Early in his career, he had spent more than a year living with the Gidjingarli people of central Arnhem Land doing ethnological work with his anthropologist wife, Betty Meehan. He guided Roberts and Smith, relative new chums, through the northern customs.

When the three published their dates of 52,000 to 61,000 years for the Malakunanja II rock shelter in *Nature* in 1990, they made headlines world wide, to the surprise of Roberts, who had come to the problem with the perspective of a geomorphologist. 'I'd been dating samples that were a quarter of a million years old,' he says. 'Our results increased the date of colonisation from 40,000 to 60,000 years, which to me seemed to be twopence-halfpenny, really. There was this big hoo hah. I was surprised there was so much interest.'

Many doubted the results from this mysterious new dating method, however, and argued that the numbers would not stand up without corroboration with old dates from other sites. In an exchange in the literature, archaeologist Sandra Bowdler accused the team of 'bullyspeak', and said of an exposition of the method for estimating the error on the dates: 'Now if we are not completely dazzled by Science, we might be able to dimly discern what they could possibly be trying to say.'

The Malakunanja results were later backed up with OSL dates on nearby Nauwalabila I of 53,000 to 60,000 years. Roberts was about to find himself in deeper controversy, however, over a site on a Dreaming track in the Northern Territory part of Kimberley region.

Headlined 'Unveiled: Outback Stonehenge that will rewrite our history', the breaking story on the Jinmium rock shelter, a big sandstone block in a place named for a female ancestral being, claimed that new ages for the site pushed the date of human colonisation of Australia back to perhaps 176,000 years ago. It sparked a media frenzy world wide. News of the sensational dates for the site in monsoon country in the territory's north-west followed a big win by Aborigines bringing a native title case in the High Court. It also came amid a paroxysm of racism in Australia. 'In the highly charged political atmosphere of the Kimberley, where Aboriginal traditional owners are trying to regain control over land from pastoralists ... the discovery is absolute dynamite,' wrote James Woodford, the *Sydney Morning Herald* reporter who broke the story in the mainstream media in September 1996.

The research team—Lesley Head, of Wollongong University, New South Wales, her archaeologist husband Richard Fullagar, then at the Australian Museum in Sydney, and dating specialist David Price, also of Wollongong University—had obtained thermoluminescence dates of between 50,000 and 75,000 years for sand associated with rock art—circular engravings, or cupules—at the site, on sands 50 kilometres from the mouth of the Keep River. The results suggested that the art, some of which was on slabs of rock that had broken off the wall of the shelter and been buried, pre-dated by tens of thousands of years France's Chauvet Cave rock paintings depicting mammoth, horses and bison. Even more spectacular were ages for artefacts, recovered from up to 160 centimetres below the surface. The ages ranged from 116,000 years to 176,000 years for a sterile level, dates that at least trebled the span of occupation of the continent. There was even talk of an 'Australian Eve'.

'As the scientists say in their paper,' Woodford wrote, 'Australia may have been originally occupied by one of the several archaic—or primitive and now extinct—human species that lived in South-east Asia. Even more controversial is the possibility that modern humans evolved from these archaic humans independently from the rest of the world during their interaction of up to 176,000 years with the Australian environment.'

Many researchers attempted to hose down the claims as the story was splashed across newspapers, including the *New York Times*, and trumpeted on television and radio news bulletins around the world. Dating experts were worried that the incident would undermine the relatively new luminescence method. The team, which had started excavations at the site in 1993, went to the media before formal publication of the research, which had been accepted by the prestigious British archaeological journal *Antiquity*.

Archaeologist Mike Morwood, now at the University of Wollongong, came out publicly as one of the reviewers of the *Antiquity* paper, revealing that he had recommended against publication on the grounds of the dating. It was an unusual step, with reviewers usually remaining anonymous, and the move created divisions that would last for years. Morwood labelled the team incautious for going public with the results before further dating work had been done.

The dating method works only if the luminescence signal in the sand grains has been 'zeroed' by sunlight or heat before the crystals are engulfed by darkness. This is time zero for the sub-atomic clock, which starts ticking as natural radioactivity begins winding up the luminescence signal again. David Price had deployed thermoluminescence dating on grains of sand from the same layers as the art and artefacts. There were plans for Roberts to use the more accurate optically stimulated method on the site, but that would be some time off. Many dating specialists, including Roberts, argued that the Jinmium samples had not been sufficiently 'bleached'. The quartz crystals had been lumbered with an 'inherited age' that would give a reading that was too old. Another sceptic was Nigel Spooner, then at the Australian National University. He later told the Australasian Archaeometry conference in Sydney, which had been organised by Fullagar and Tuniz, that evidence for insufficient bleaching lay in the Jinmium team's own data, in the 'glow curve'—a plot of the light emission from the sample against temperature. He published his analysis in *Antiquity*. But the team, for now, was standing by its results. Fullagar told *The Canberra Times* that he respected Spooner but, 'We're not going to suddenly say that the dates are wrong because of the reinterpretation of somebody eyeballing graphs and figures and things.'

The topic was still hot when *Antiquity* published the Fullagar/Head/Price paper in December 1996. In his editorial, Christopher Chippindale defended his decision to accept the paper in the face of a negative review, and to sanction release of the story to the media before publication in the journal. Claims that the results conflicted with the out of Africa theory were no cause to hold back either, he said. With estimates for the emergence of modern humans in Africa ranging from 100,000 to 200,000 years ago, the Jinmium results did not upset the model. 'I see no cause for us to have held back on Jinmium for fear that some colleagues, or some fast-thinking journalists, might rush to decide it must torpedo the out-of-Africa model of *sapiens* origin. It does not, as the authors say. (And they did rush.)'

The paper said:

Age estimates for the earliest modern hominids in the world range from about 100,000 to 200,000 years ... The chronology and classification of East Asian hominids at this time is patchy, though archaic forms (Ngandong in Java) are thought to date to about 100,000 years ... it is not impossible for fully modern hominids to be in South-east Asia at this time, though evidence of humans in Australia at this earlier time (compared with a time-frame of around 60,000 years from the dates for lowest strata in Arnhem Land sites) increases the chances of colonisation by archaic humans. There was an extended period of very low sea level prior to 135,000 years and several periods of low sea level during the Last Interglacial, which may have enabled easier water crossings by humans from Southeastern Asia to Greater Australia.

Multiregionalists seized on the results. 'Any date beyond 120,000 years makes it difficult for the out of Africa theorists,' the ANU's Alan Thorne was quoted by *New Scientist* as saying. 'Their dates would suggest that if modern humans evolved in Africa, they must have invented the bicycle at the same time so they could cycle around to catch the first rafts to Australia.'

Roberts redated the site in 1998, deploying optically stimulated luminescence. The OSL signal can be reset within a few seconds or minutes of exposure to sunlight, against the hours or days it takes to zero the thermoluminescence clock. The reasons lie in the realms of quantum physics. Both methods require the careful measurement of the site's radiation environment. OSL also enables the analysis of single grains, allowing the direct assessment of the problem of 'inherited age'. The Roberts team, which included luminescence dating expert Jon Olley and statisticians Rex Galbraith and Geoff Laslett, tested 1,000 individual sand grains, and was able to exclude from the analysis crystals that had not been zeroed. The luminescence results, backed up by radiocarbon dates on charcoal, obtained by Tuniz's group at the Australian Nuclear Science and Technology Organisation, showed that the site was less than 10,000 years old. The Roberts team, including Fullagar, reported its results in *Nature*. The results squared with the out of Africa model; the luminescence dating method had been redeemed.

Attention shifted to the Malakunanja and Nauwalabila sites in Arnhem Land. Malakunanja had been dated by thermoluminescence, the method used by the Jinmium team, and this threw the reliability of the date into question. Roberts and colleagues redated the site by OSL on single grains, firming up their 60,000-year chronology. It was not enough to silence all critics.

Soon after the Jinmium imbroglio, attention returned south to the most closely studied, and politically charged, site in Australia—the Willandra Lakes. Research there had been sporadic since the big excavations of the 1970s. Even seemingly innocuous pursuits, such as the dating of shell middens, took on a political aspect, as all research was conflated with the sensitive issue of the removal of human remains. In 1989, Aborigines and researchers, meeting at the Willandra Research Publication Workshop, had signed an agreement—'The Mungo Statement: Towards a Reconciliation'—on the future of research at the site. 'It was decided to embark on a course of reconciliation between archaeologists and Aborigines,' the statement said. 'It was recognised that Aboriginal people must have the final say whether research was done and what it might be.'

Workshop delegates proposed an Aboriginal research committee to oversee and ratify research programs 'dealing with the way people lived in the past and the kind of land they lived in'. The committee could initiate its own research programs and seek funding for them. Alan Thorne agreed that skeletal remains under his care at the ANU would be returned to the Willandra. As a first step, a second lock would be fixed to the safe in which they were stored, with representatives of the Aboriginal community holding the key. The meeting also called for a keeping place at Mungo National Park to house remains, with one key held by indigenous representatives and the other by the scientific community. Several scientists were unsettled by the statement, fearing it smacked of censorship and suppression of research. Some steered clear of the Willandra altogether because they viewed doing research there as just too hard.

While the politics was daunting, so was the science. The first dating work was done in the 1960s during the second decade of radiocarbon dating, when the site was a testbed for the method, the subtleties of which continue to vex the best minds in the field. Two big research programs, one in the late 1980s and the other a decade later, would focus new techniques on the problem, but scientists still argue about the results. Some say the original radiocarbon record, compiled during the early days of research, is firm. Others disagree, and have pinned their hopes on luminescence and other dating techniques.

3 Mungo Lady gets date

'I'll just take a short walk to get a feel for the stratigraphy.'

Jim Bowler, a tall, reflective man rarely seen without his Akubra hat, is 'interviewing the landscape' at Lake Mungo for colleagues wanting an idea of the age of a site before going to the expense of dating it precisely. What looks to most like an amorphous mass of sediments resolves itself in his eyes into a geological sequence with the clarity of a textbook stratigraphic diagram encoding the response of the landscape to the forces of climate and people. He is best known for the discoveries of Mungo Lady and Mungo Man when he was a research fellow at the Australian National University. But this is his real talent—and 'flying kites', one of his latest being a reinterpretation of the geological record that puts the Antarctic ice sheet centre stage in global climate change. Bowler's kites have a tendency to become orthodoxy.

After beginning his work, Bowler progressed upward through the geological strata to focus eventually on the Pleistocene. His studies of the stratigraphy and palaeoclimate of the Willandra Lakes began in 1967, following a suggestion by geologist Joe Jennings, an associate professor in the Department of Biogeography and Geomorphology. Jennings had spotted the dry lakebeds on a flight from Broken Hill to Melbourne. Bowler had previously worked on Victorian crater lakes that still held water, and whose sediments could be read as a rain gauge of the past. His 1971 *Nature* paper, with Tatsuji Hamada, of the RIKEN radiocarbon lab in Tokyo, on water levels in Lake Keilambete is a classic in palaeohydrology. By the time he got to 'Bio and Geo', the neighbouring Prehistory Department had commissioned a new radiocarbon lab. Both departments were housed in the then Research School of Pacific Studies, already a formidable force in prehistory studies.

The Willandra Lakes were absent from most maps and were unknown to science, although sheep farming had been established in the area by the mid-nineteenth century, and there were still working stations when Bowler turned up. Several other dry lakebeds in the semi-arid, sparsely vegetated land between the Murray and Darling rivers are coloured an enticing light blue on tourist maps, to the bewilderment of adventurers in four-wheel drives with boats on top.

Bowler mapped the region, finding that the lakes were part of a relict drainage system on Willandra Creek, a tributary of the Lachlan River which had formerly reached the Murray. He named the biggest lakes after local sheep stations: Mulurulu, Garnpung, Leaghur, Mungo and Outer Arumpo (a small lake inside the big one was already called 'Arumpo'). Lake Mungo, at the heart of the system, got up to 8 metres deep when its source to the north, Lake Leaghur, overflowed. Bowler looked for exposures—sites where erosion has revealed deeply buried layers.

On July 5, 1968, he spotted a pile of carbonate-encrusted, burnt bones weathering out of the lunette at the southern end of Lake Mungo. He marked the location with a star picket and photographed the bones, but on his return to Canberra found it difficult to arouse much interest among the archaeologists. Prehistory was booming, and resources were limited. Writing much later in the journal *Australian Archaeology*, John Mulvaney recalled:

> Jim Bowler reported the discovery of artefacts and hearths on the Pleistocene shores of Lake Mungo to myself and others during 1968. One hearth in particular contained bones which he believed might belong to an extinct marsupial. For various reasons archaeologists were slow to respond.

Mulvaney made a pitch to John Barnes, then chair of the Anthropology Department, and Peter Grimshaw, the research school's business manager, for funds for fieldwork at Mungo. It had all the right buzzwords:

> Mr Bowler (Biogeography and Geomorphology) has located several archaeological sites of Pleistocene age in the western Riverina, on the Lachlan River. There are some indications that occupation may be considerably older than 25,000 years, the present oldest known age of Aboriginal habitation in Australia. There is also a likelihood of finding remains in association with bones of extinct giant marsupial fauna.

A preliminary budget estimate included fuel for the university's VW Kombi for a 1,150-mile (1,850-kilometre) round trip to Mungo, accommodation for three people (two nights at the Balranald Motel, one night camping), and meals. The total amount requested was $94.

Mulvaney wryly recalled: 'As the hearth actually contained the Pleistocene human cremation ("Mungo Lady"), and Harry Allen completed his PhD in the region, the meagre expenditure proved a remarkably productive investment that ensured the Willandra Lakes a World Heritage registration by 1981.'

The request granted, archaeologists Mulvaney, Rhys Jones, Con Key and Harry Allen, and a separate party of earth scientists, met Bowler at the site in March 1969. Two burnt skull and jawbone fragments looked human, and the white carbonate-cemented sediments entombing them attested to their great antiquity. Then Jones found a human tooth. Geologist Keith Crook remembers Jones holding aloft the diagnostic tooth and dancing around the burial. This was no extinct marsupial—it was a Pleistocene human burial. The team had not come prepared for an excavation, but something had to be done because the remains were already scattered and vulnerable to wind, rain and trampling by sheep. A fierce thunderstorm was brewing. Bowler had warned the party of the transitory nature of objects in surface exposures at Lake Mungo. The archaeologists collected the dozens of loose bone fragments, as well as some still embedded in carbonate lumps, and packed them all into John Mulvaney's suitcase for transport back to Canberra. The bones arrived safely, and were soon confirmed as human by Alan Thorne and zoologist John Calaby. Mulvaney's suitcase is now housed in the National Museum of Australia.

Soon after, Jones and Allen returned to the site with Bowler, the archaeologists collecting small flaked tools made of silcrete, a rock forged from sand grains deposited by the sea which had encroached on the interior millions of years before. The tools did not fit the European system of Lower, Middle and Upper Palaeolithic, or the African Early, Middle and Later Stone Age, and the Australian archaeologists wanted to make this clear. Many of the hearths with silcrete blades, scrapers and choppers at Mungo were scattered in a zone tracing the former lake shoreline, about 20 metres from the water's edge. Jones and Allen labelled the technology the 'Australian core tool and scraper tradition' when they published their results in *World Archaeology*.

Ancient hearths stood out as dark patches in the white quartz sediments of what Bowler labelled the 'Mungo stratigraphic unit', formed during a lake-full stage when wind had blown sand onto the lunettes from beaches. Already, the group suspected that the site was going to be of high international significance, and its mysteries were gripping. The biggest was how the ancients disposed of the body of the diminutive lady, who was just shy of 1.5 metres tall and still in her twenties when she died. Her people cremated her, removing her skeleton from the coals and smashing it, especially the head, before returning it to the fire and covering it with sand. Her remains left no forensic clues to the cause of her death.

Soon after Mungo Lady's discovery, Harry Allen surveyed four of the lakes, Mulurulu, Garnpung, Leaghur and Mungo. It was Australia's first large-scale regional archaeological survey: most previous work had been confined to single sites like rock shelters. The title of his PhD thesis—'Where the crow flies backwards'—was a statement on modern environmental conditions in the Willandra.

Animal bones found in the hearths and middens were mostly from small marsupials. Birds, lizards, fish, shellfish, eggshells and yabbies (freshwater crayfish) were also on the menu. Allen compared fish otoliths and vertebrae with those from golden perch (*Macquaria ambigua*) bought at a Mildura fish 'n' chip shop. The otolith, or 'ear stone', is made of calcium carbonate, the stuff of shell, and similarly in the 'aragonite' crystal form. It is suspended in a fluid and forms part of the fish's balance and orientation system. These dense, translucent, ear-shaped discs, up to 20 millimetres long, can survive in the geological record for hundreds of thousands of years. Some of the golden perch caught long ago in the Willandra Lakes were whoppers, almost a metre long, and growth rings in the otoliths showed the fish were up to 50 years old at death. None of the hearths and middens had bones from big red or grey kangaroos, but the pit ovens in the Mungo and Outer Arumpo lunettes were probably used for cooking large animals. The archaeologist, Allen, called palaeoanthropologist Alan Thorne out whenever he found a burial, and looked for material suitable for radiocarbon dating to flesh out the lifestyle of the ancients.

***

The radiocarbon revolution was well under way when the early work at Mungo began, with dates from the famous Cro-Magnon burials in France coming in at about 34,000 years. Until the 1960s, Australian prehistory was thought to have been brief and uneventful. Aborigines had probably been in Australia for less than 10,000 years, and hadn't done much. Mulvaney's excavations at Kenniff Cave in Queensland changed all that. And precisely one hundred years after the Cro-Magnon discoveries, Jim Bowler found Mungo Lady—who was possibly older—although Bowler had to wait until 1969 for formal identification of the remains as human.

Also in 1969, Neil Armstrong collected moon rock for analysis back on Earth. And a meteorite crash-landed near the Victorian town of Murchison, south-east of Mungo. Chemical analysis of the meteorite revealed eight of the amino acids in proteins, and three of the four bases that make up DNA, fuelling speculation that Earth had been seeded with life from another planet. Nobody had yet considered the possibility of extracting DNA from ancient bones, or using modern DNA to test theories of human evolution.

The 'Man the Hunter' conference had been held in Chicago in 1966. It put the study of modern hunter-gatherer societies on the academic agenda, but it also aroused controversy. Papers presented at the conference implied that men did most of the hunting in prehistoric societies, and it was this 'men's work' that had driven the evolution of the brain; women had simply ridden in a genetic slipstream. The fallout continues.

And every hip household had a pressure cooker—or almost every. The Polach family in Canberra didn't, because Henry Polach had taken it in to work at the ANU's new radiocarbon laboratory. The larger-than-life Czech refugee, who dominated a room just by walking into it, had fought with the Resistance during World War II, making bombs using the chemistry skills he had picked up during a since-abandoned medical degree. In Canberra, he had commandeered the kitchen appliance in the hope of developing a better method to extract collagen, the most abundant protein in the body, from Mungo Lady's bones to date them directly. His resourcefulness and 'bucket chemistry' belied a growing body of expertise and an arsenal of hi-tech instrumentation that would put Australia in the forefront of quantitative dating, a field centred on the so-called 'hard sciences' of physics and chemistry.

There was no good precedent for dating human burials in Australia. The stratigraphic context of many skeletons found earlier was unknown, so it was impossible to date them indirectly through charcoal or shell from sediments associated with them. And scientists were sceptical about direct radiocarbon dates on bone. Even Willard Libby, the American chemist and Manhattan Project veteran who won the Nobel Prize in 1960 for inventing the radiocarbon method, was doubtful. Bones that had been buried for a long time are often porous, with little of the original organic material left.

Radiocarbon dating is based on the decay of the radioactive isotope of carbon—carbon-14—formed in trace amounts in the atmosphere. Cosmic rays, mainly protons blasted into space by supernovae, bombard nitrogen and oxygen atoms in the stratosphere, up to 80 kilometres above the Earth's surface, forcing them to eject neutrons. These electrically neutral particles bounce off atoms of gas, with some slowing down enough to trigger atmospheric nuclear reactions. When slow neutrons collide with nitrogen atoms, they form carbon-14, with most of the action happening about 15 kilometres above the Earth.

Radiocarbon reaches an equilibrium value in the atmosphere

of about one atom per thousand billion atoms of the much more common stable, or non-radioactive, isotopes of the element of life: carbon-12 accounts for 99 per cent of the element, with carbon-13 making up about 1 per cent. The production rate of radiocarbon, and its atmospheric concentration at any given time, depend in part on the strength of the Earth's magnetic field, which shields the planet from cosmic rays, and on fluctuations in the sun's magnetic field. Carbon is distributed between several reservoirs, with about 93 per cent in the hydrosphere (mainly the oceans), 5 per cent in the biosphere, and 2 per cent in the atmosphere. It is exchanged between these reservoirs via mechanisms that can influence the carbon-14 to carbon-12 isotopic ratio. For example, the upwelling of older waters from the deep ocean (where carbon-14 had the time to decay away) can decrease radiocarbon concentrations in living fish and shellfish.

Carbon-14 bonds with oxygen to form carbon dioxide, and enters the food chain via respiration and photosynthesis. Its concentration in living organisms is about the same as in the atmosphere, about one part per thousand billion. After tissue is formed, the radiocarbon concentration starts decreasing at a known rate through radioactive decay to nitrogen-14. It takes 5,730 years for half of the radiocarbon atoms originally present to disintegrate, and the residual carbon-14 concentration in the sample reveals when it was formed.

But very old samples have so little radiocarbon left that the limits of detection are reached. After 50,000 years, the radiocarbon concentration becomes one part per thousand trillion. And contamination from younger carbon invading the sample from the environment after the plant or animal has died or during sample preparation in the laboratory can swamp this low concentration of atoms, knocking thousands of years off the true age. Together, these problems throw up a 'radiocarbon barrier', likened by ANU geologist John Chappell to an event horizon beyond which all results fall into a radiocarbon black hole. It's a signal-to-noise problem. When Polach was working on the Mungo samples, the barrier stood at about 40,000 years, a number that crops up regularly in Australian archaeology. It was imperative to collect as much datable material as possible, and preferably several different materials.

Bowler had already sent Polach charcoal from fireplaces and shells from middens, and samples from the stratum bearing the bones indirectly placed the Mungo Lady burial between 29,000 and 36,000 years old. However, some of the charcoal dates were coming in younger than shell dates from the same layers, and the pressure was on to get direct dates on the bones.

The calcium phosphate mineral apatite—the matrix of bone—has a small carbonate component, which like all carbonates is prone to contamination. Collagen, the rope-like protein that binds the apatite crystals together, is the most reliable source of organic carbon for dating bones, but it degrades quickly. The ANU radiocarbon lab had just a few precious fragments of bone from Mungo Lady to work on.

Polach was interested in the new 'Longin method' for isolating and purifying collagen, named after French scientist Robert Longin, who published it in a 1971 *Nature* paper. Longin found that collagen could be converted to its soluble form, gelatin, by gently warming it in very weak acid. This was a novel application of animal glue and soup recipes thousands of years old, and Polach, an accomplished cook, thought he'd go one better by using a pressure cooker to speed up the process. Practising on animal bones, he blew up the appliance three times. His last words on the subject were, 'Bloody bones, never again!'

He reverted to a more conservative method to extract the organic material, which he presumed had some collagen. He burnt this fraction to convert it to benzene for liquid scintillation counting, then one of the two main methods for measuring radiocarbon concentrations. He loaded the sample into the ANU's new scintillation counter, an instrument the size of a household freezer. He had mixed the sample with a fluorescent material that emitted a flash of light when it absorbed energy from an electron, or beta particle, liberated in the decay of a carbon-14 atom. A photomultiplier converted the flashes of light into electrical signals, which revealed the amount of radiocarbon left in the sample, and therefore its age. For very old samples, it could take days or weeks to get enough counts for a precise figure.

Many were disappointed with the result, reported in *Nature* in 1972, that Mungo Lady had been cremated about 29,000 years ago. The number raised more questions than it answered. Accompanying

Polach's paper was another by Michael Barbetti and Harry Allen, reporting radiocarbon dates of hearths and middens indicating occupation back to 36,000 years. In a commentary in the journal, Mulvaney said the work suggested that modern man had 'just possibly voyaged to Australia before he reached the New World or even Europe'.

The denudation captured by Russell Drysdale in his painting of the Walls of China looked different through the eyes of an archaeologist. In his *Nature* commentary, Mulvaney wrote that Drysdale had painted 'a country stricken by drought, and his vision is of a melancholy landscape, riven by erosion and studded with contorted remnants of dunes and grotesque tree stumps blasted by sand'. But to Mulvaney, the denudation of the lunette had provided a 'geomorphic and archaeological bonus'.

***

In the early 1970s, an Australian archaeological journal ran a short paper by Jim Stockton, of Monaro Road Constructions in Canberra. It said: 'The backhoe is potentially an extremely valuable tool for archaeology.' For coarse stripping of overburden, the bulldozer and scraper combination was best, and the Caterpillar D8 was 'an extremely beautiful machine'. For the delicate work of fine stripping 'where precision is necessary and damage must be minimal, a grader is most effective'.

Stockton was pitching to a generation of researchers who didn't muck around. Bowler, who came from a potato-farming family, bought a bright yellow tractor, with backhoe and front blade attachments, to help Mulvaney and Wilfred Shawcross in their archaeological excavations at Mungo in the mid-1970s. He had encouraged Mulvaney to put a joint research proposal to the Australian Institute of Aboriginal Studies. They requested money for a geological technician, a tractor, and the chartering of a light plane for aerial photography. John Magee, a recent geology graduate at ANU, got the job as field and lab technician. Large grants were rare then, but the Pleistocene radiocarbon dates for occupation of the Willandra, the oldest in Australia, ensured the team got the funding.

Much red tape would have to be cut through before work could start on what would be Australia's first tentative plunge into 'big archaeology'. The National Parks and Wildlife Service had site preservation to consider, and rumours abounded that the ANU team would destroy the site with their heavy earthmoving equipment. The archaeologists had to write an environmental impact statement, and were finally given a permit on the condition that they back-filled their trenches. There were no such restrictions on Bowler because he was a geologist, and geologists were allowed to dig holes just about anywhere in minerals-rich Australia. Bowler and Magee filled in details on the map of the Mungo lunette and dry lakebed.

Mulvaney's trench was set back in a less eroded part of the lunette, with Shawcross's nearer the Mungo Lady burial. ANU archaeologist Isabel McBryde dug hearths and middens on the Outer Arumpo lunette, with Bowler's tractor deployed to help remove the overburden. She later estimated the shellfish meat yield from one midden at more than 40 kilograms.

Even in those early days, there were hints of greater antiquity of the site. Mulvaney, for one, had obtained a non-finite date—beyond the 40,000-year radiocarbon limit—on a tiny charcoal sample from near the bottom of his trench. He never published the date—the sample was too small and too old for the available technology—but everyone talked about the result because it was older than the Cro-Magnon burials in France. Surely modern people could not have reached the antipodes before they got to Europe!

However, the age of the site promising to answer one of the biggest questions in prehistory was still proving elusive. In 1987, archaeologist Peter Clark listed 150 radiocarbon dates that had accumulated on Willandra bones, charcoal and shells, but the results were contradictory. Dates on some charcoal samples were still coming in younger than midden shells, and it was the shell dates that were a better fit to Bowler's stratigraphy. This contradicted the conventional wisdom that charcoal was top dog and freshwater shell unreliable.

***

Under a microscope, charred plants look like fine black lace. The cell walls remain intact, forming an exquisite lattice, but the interior of the cells is gone—destroyed in the blaze. The remaining highly porous structure gives charcoal its absorbent properties, and this opens up the possibility for contamination by younger carbon. Charcoal from ancient hearths is the material usually dated by the radiocarbon method, but any unburnt wood remaining in the charcoal particles slowly breaks down. Large 'macromolecules', such as cellulose and lignin, are chopped into smaller molecules, which are consumed and recycled into other compounds by soil microbes. Countless generations of microbes mediate an interchange of carbon between the charcoal and its environment over thousands of years.

Some of the carbon in the unburnt firewood is redeposited within ancient fireplaces as humic acids, a ubiquitous brown-black gunge that has been the bane of radiocarbon chemists. The material gets darker, and is sometimes the only organic material left in a fireplace—like the Cheshire Cat's grin after the cat has disappeared—and it sticks to sand and clay particles, forming black lumps. What looks like charcoal in what was obviously a fireplace is the decomposed remains of the Cheshire Cat's last meal of red herring.

Dating specialists can dissolve out the humic acids when they pre-treat the samples, but in the late 1980s, when Richard Gillespie put the remaining material from Willandra samples under a microscope, he found that many lacked the lace-like structure of charcoal. There had been no charcoal there in the first place. Perhaps the ancients had used grass or dung rather than wood as the fuel to cook their meals.

Meanwhile, another problem related to a drawback in radiocarbon dating itself, and Australians had been slow to pick up on it.

***

Radiocarbon dating revolutionised prehistory, anchoring in time events previously drifting as relative chronologies based on stratigraphic sequences. In the early 1950s, radiocarbon dates coming in from important Egyptian sites matched historical dates closely, and this seemed to vindicate the method. Stone Age, Bronze Age and Iron Age, a stratigraphic approach developed by nineteenth-century Danish archaeologists, could now be discussed in years. Dates from widely separated sites could fill in the details of demographic and technological changes far beyond textual records.

But something nasty lurked in the background.

As the number of radiocarbon dates increased and reached back further in time, discrepancies emerged. Dates for the Egyptian Old Kingdom seemed to be coming in too young compared with written sources, some by as much as 800 years. Hessel de Vries at the Groningen laboratory in Holland showed that the radiocarbon chronology disagreed with highly accurate tree ring chronology, or dendrochronology. The finding sent shock waves through the scientific community: the radiocarbon method itself was at stake. Creationists seized on the discrepancy as support for the 4004 BC date for the creation of the Earth reckoned by seventeenth-century Irish archbishop James Ussher.

Libby had tested his new dating method on samples whose ages were known from history, including wood from Egyptian pharaoh Sneferu's pyramid thought to be about 4,500 years old. He also checked it against the tree ring record. The age of his oldest wood sample was known exactly from the number of growth rings. It was from the Centennial Stump, a huge redwood chopped down in 1874—something to do with US celebrations of nationhood—and Libby's sample consisted of rings that had grown between 938 and 1031 BC. When he published his 'curve of knowns' in 1949, Libby alluded to a weakness in the method—the assumption that cosmic radiation flux, and therefore the atmospheric radiocarbon concentration, had always been constant.

The dendrochronologists who challenged this assumption came from a scientific lineage started in the early twentieth century by Andrew Douglass at the Tree Ring Laboratory in Tucson, Arizona. A later resident there can probably be dated in time by car buffs from his confession at a radiocarbon meeting that he, and others in his business, 'wouldn't know an atom of carbon-14 from a BMW 2002'. The arcane science of tree rings proved to be the salvation of radiocarbon dating, clearing the way to calibrating radiocarbon dates by factoring in the fluctuating atmospheric radiocarbon levels. The difference between a tree ring's measured radiocarbon age and its exactly known tree ring age could be plotted on a graph. For any given radiocarbon age, the growth ring with the same radiocarbon age revealed the corresponding calendar age.

Hans Suess, of the La Jolla lab in California, published one of the first calibration curves in 1970. He had accumulated hundreds of radiocarbon measurements on tree rings, plotted them against the tree rings' known ages, and used what he called 'cosmic schwung' to draw by eye a curve best fitting the points on his graph. The smooth curve based on meagre early results now had 'wiggles'.

The internationally recognised tree ring calibration curve of today, with its short timescale wiggles superimposed on large timescale peaks and troughs, stretches back about 12,000 years. The wiggles, with periods of 11 and 22 years, correspond to the sun's magnetic variability, the major cause of carbon-14 fluctuations over the past 15,000 years. The pattern of peaks and troughs, with a period of 8,000 years, reflects the Earth's magnetic field changes. The drawing of the calibration curve has been painstaking—counting the rings and measuring their width in thousands of living and long-dead trees to create an overlapping sequence, processing closely spaced wood samples to remove carbon contamination, and measuring the radiocarbon content to high precision. One 7,000-year section of the curve, produced from 20-year blocks of bog oak trees by a team at the Palaeoecology Centre at Belfast, took 15 person-years of dedicated work. The complete tree ring calibration curve has now been cross-checked at multiple labs.

There are prospects for extending the tree ring record into the Pleistocene. Celery-top, King Billy and Huon pines from Tasmania have radiocarbon ages up to 17,000 years. And buried kauri logs being excavated in New Zealand could cover the complete 50,000-year range of radiocarbon dating. Other natural systems with annual or seasonal layers, including coral, stalagmites and stalactites (speleothems), varves (layers of lake sediments), polar ice and marine sediments, have extended the curve back to about 50,000 years. Coral and speleothem growth bands are dated independently using a method based on the radioactive decay of naturally occurring uranium to its daughter products, while varves—deposited by melting snow and ice—can, like tree rings, be counted directly. These techniques carry uncertainties of about 2,000 years for Pleistocene ages, against an uncertainty of only 20 years for the dendrochronology-based calibration in the

Holocene.

There is no consensus on the exact shape of the calibration curve beyond 26,000 calendar years ago; measurements from different sites disagree, although a general trend is evident. This book uses only calibrated radiocarbon dates calculated with the University of Cologne program Calpal-2007~Hulu~. Calibrated radiocarbon ages are directly comparable with ice core ages and results from uranium-series and OSL dating methods. Because we use only calibrated

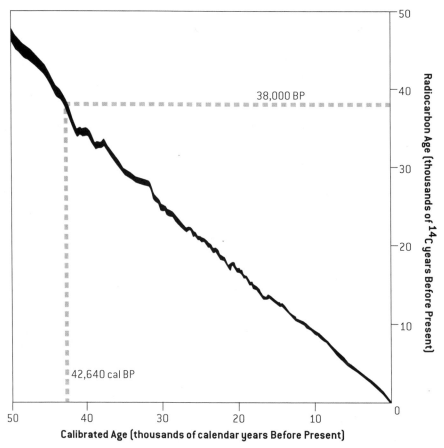

Calibration curve to convert radiocarbon ages to calendar ages, CalPal2007~Hulu~ from the Cologne Radiocarbon Calibration and Palaeoclimate Research Package (available from www.calpal.de/). Radiocarbon dates require calibration because the amount of carbon-14 in the atmosphere has changed over time, and calibration is essential for comparison of radiocarbon dates with the results from other dating methods using calendar years.

radiocarbon ages (for example 42,640 cal BP, as in the figure) the numbers for well-known samples and sites will differ from those originally published.

* * *

Back in Australia in 1987, following stints in radiocarbon labs at the universities of Oxford and Arizona, Richard Gillespie started work at ANU with a creed of 'better chemistry'. He began redating some of the Willandra Lakes human remains, along with charcoal and shells from Isabel McBryde's excavations. He could not get any protein out of the Mungo Lady remains, finding instead that more than 90 per cent of the carbon in the bones was humic acid; dates on this component came out the same as Polach's result of 29,000 years. Other burnt bones from the Willandra were similar, and only two out of 56 skeletons analysed had any collagen. One came in at 5,700 years old, while the other, intriguingly, dated from the second half of the twentieth century. In a review of the radiocarbon dates, Gillespie eliminated results for suspect charcoal and bone samples, and calibrated the rest. The radiocarbon dates from the layer bearing the Mungo Lady burial and the almost intact Mungo Man skeleton found by Bowler in 1974 converged on 40,000 calendar years. Perhaps Mulvaney's figure was right all along.

In the late 1990s another attempt would be made to date the oldest burials, through a combination of methods. One of them, electron spin resonance (ESR) dating, has been used by the ANU's Rainer Grün to date some of the world's most important fossils, including Neanderthals and early modern humans. ESR dating of tooth enamel uses samples so small it is virtually non-destructive. And as a direct dating method, it sidesteps questions about the association of the datable material with the object of interest.

The magnetic field generated by Grün's time machine stops your watch if you get too close. Like his colleagues, he is burdened with a dating method few grasp. His method has a reach of 2 million years. Invented in 1967, ESR dating was first applied in 1975, but is still being developed. It is rooted in the strange realm of quantum physics. Grün writes papers that 'no-one understands', and which don't become any more comprehensible when he gives people translations of them in his native German. 'You put a sample into an ESR machine and get a number out,' he has said wearily when asked to explain how his method really works.

Like luminescence dating, the ESR technique exploits the impact of natural radioactive elements, mainly uranium, and cosmic rays on the electron structure of crystals. Unpaired, spinning electrons left behind when their partners get trapped in crystal defects are sub-atomic magnets. Grün's magnet aligns their spin direction. By hitting the sample with microwaves, the electrons are forced to reverse their spin direction. The amount of microwave energy absorbed in the backflip reflects the number of lone electrons and the age of the sample. However, a good estimate of the radiation dose the sample received every year it was buried in the ground is critical, and this is where the method gets tricky. Teeth in living animals have little uranium but buried teeth take the element up from ground water. At issue is whether the uranium entered the tooth at once, soon after the animal died, or accumulated gradually over the millennia. Grün would later find a way to combat the problem, but not until after his analysis of the Mungo Man remains.

In 1999, Alan Thorne, Grün and colleagues published ages of around 62,000 years ago for the Mungo Man skeleton based on ESR and other dating methods—OSL, conducted by Nigel Spooner, and uranium-series, based on the rate of decay of uranium to its daughter products.

The electron spin resonance method gave two age estimates, of 63,000 and 78,000 years, and OSL dating produced two more, of 58,000 and 63,000 years. Uranium-series ages measured by gamma spectrometry on the skull ranged from 60,000 to 74,000 years. The 305-gram partial skull was put in a chamber shielded from the outside with lead and fitted with detectors to count, over two 50-day periods, the high-energy gamma radiation emitted when uranium and its daughter products in the skull decayed. Other uranium-series measurements on long bone fragments yielded a further four numbers ranging from 54,000 to 70,000 years. The team combined the results, getting an age range of 56,000 to 68,000 years. With Lake Mungo well inland and 2,700 kilometres from the present north-western coast of Australia, initial colonisation of the continent could have been much earlier, the team said in its paper, published in the *Journal of Human Evolution*.

The results suggested that the 'gracile', or slight, Willandra Lakes people pre-dated by 40,000 years the more thickset 'robust' individuals from elsewhere in Australia. (Neanderthals and early members of the genus *Homo* were robust, but humans today are at the gracile end of the range.) The paper was measured in its discussion on the significance of the dates for debates on human evolution. Multiregionalists, it said, would have to explain why 'gracility appears first in Australia when Indonesia, the nearest source area for migrants, had a long history of robusticity'. Proponents of the out of Africa model, it argued, would have to explain how robusticity developed relatively quickly in Australia when 'gradual gracialisation characterises human evolution over at least the last 200,000 years'.

The comments echoed those made by Mulvaney in his 1972 overview article. He noted that although Thorne suggested in *Nature* the same year that skeletons from Kow Swamp in north-western Victoria exhibited 'archaic' traits, those burials were '... some 15,000 years more recent than the modern cranial form at Mungo, and no topographic or environmental barrier separated these populations'.

A separate team led by Bowler and including both Spooner and Roberts redated the site in 2003, getting an OSL age range of 38,000 to 42,000 years for the Mungo Man and Mungo Lady burials. Sediments containing the oldest stone tools in the 1970s Shawcross excavation came in at between 46,000 and 50,000 years old.

That team argued that the OSL samples from the earlier study, which had been collected 300 metres from the Mungo Man burial, had been from a lower, and therefore older, stratum than that bearing the skeleton. And the uranium-series and ESR dates were compromised by problems associated with variable uptake and loss of the radioactive elements in and around the skeleton. In the later study, sediments directly above and below the burial site were sampled for OSL dating from 'Bowler holes' scattered across the eroded lunette. The Shawcross trench was redug to sample the Mungo unit stratigraphy directly. Later, in 2006, Jon Olley, then of Australia's premier science agency, the Commonwealth Scientific and Industrial Research Organisation (CSIRO), and colleagues obtained an OSL age of 41,000 years on single grains of quartz sand extracted from a resin-impregnated sample of the Mungo Man burial sands that Bowler had collected during the 1974 excavation to study the sediment composition.

Some scholars were unsettled by the paradox of a 62,000-year-old skeleton buried in 40,000-year-old sediments. Others said the new dates supported the lower end of the 50,000 to 60,000 year OSL age range for the first landfall from the Northern Territory sites, Malakunanja and Nauwalabila, while challenging the 40,000-year figure still clung to by some researchers. Roberts said it was premature to discount a 55,000 year date of colonisation for Australia, with first settlement up north being followed by a peopling of the continent over the next few thousand years. Grün's combined ESR–U-series dates, and the OSL dates of Roberts and Spooner, remain in play.

While these new dating methods were progressing beyond the radiocarbon event horizon, that venerable method was not standing still. Particle accelerators were now routinely used for carbon-14 measurement, and there were further advances in decontamination chemistry.

**

Travelling at 10 per cent of the speed of light, a beam of carbon ions from a charcoal sample excavated from a limestone cave in Western Australia was powered 22 metres through 'the tank' of the ANU's particle accelerator. A magnet powerful enough to lift a car bent the beam, with the angle of deflection dependent on the mass of the ions. The particles were channelled through magnetic and electric fields sorting them according to their mass and charge, finally penetrating the thin window of a sensitive detector capable of counting single ions.

Nuclear physicist Keith Fifield has been doing radiocarbon dating on the particle accelerator since the late 1980s, when he caught the prehistory bug. The instrument was commissioned in 1975 but has had so many modifications since that it bears little resemblance to the machine that was delivered. Along with the ANTARES accelerator at the Australian Nuclear Science and Technology Organisation, the accelerator has kept Australia up to speed on radiocarbon dating, and is also used for fundamental physics research. Accelerator mass spectrometry (AMS) has revolutionised radiocarbon dating, replacing the liquid scintillation or gas counters once used in most laboratories around the world.

AMS delivers two major advantages over traditional radiometric counting—the amount of carbon required is reduced thousands of times, from many grams to less than one milligram, and carbon-14 measurements are done hundreds of times faster. As a bonus, the small samples can be batch-processed through the decontamination chemistry. The protein keratin from a single human hair, weighing perhaps 1 milligram, contains about 50 million carbon-14 atoms. Measuring the radiocarbon concentration to 1 per cent uncertainty requires the counting of 10,000 atoms. That would take a year with liquid scintillation. AMS can do it in one minute.

The ANU's accelerator tubes are in a vertical beamline in a tower as high as a 10-storey building, one of the tallest structures in Canberra. The sample is introduced at the top of the tower and its carbon atoms are ionised and accelerated in two stages by millions of volts to the detector. The differential expansion of the tower as the sun strikes it at different points throughout the day causes tiny movements in the beamline, for which the scientists have to compensate. The accelerator has other idiosyncrasies, too. 'It always throws up little issues,' says Fifield. 'It's extremely reliable though. Those of us who use it a lot play it like a musical instrument.' The ions are travelling so fast that relativistic effects, albeit weak ones, force the scientists to adjust the calibration of the field induced by the 30-tonne magnet to compensate. 'You need a slightly stronger field for these energetic particles than you might have initially calculated,' says Fifield. The detector is a hi-tech version of the ionisation chamber used by Ernest Rutherford in his early nuclear physics experiments a century ago. 'Rutherford would have felt at home with this detector,' says Fifield. 'But it's rather refined compared with what he would have used. It gives us multiple measurements of the energy loss of the ion as it slows down in the gas. It's important in particle identification in AMS.'

The carbon-14 is counted for 10 minutes, against the days or weeks it took Polach to get his scintillation counts. On high-quality samples, the machine can get back to about 65,000 years. The limit in that age range is contamination, and the accelerator routinely gets numbers that far back on geological samples, which are often so big that modern contaminants are trivial compared with the real stuff. Beyond that, the physical limit comes into play—the minimum number of atoms that can be counted. 'At 50,000 years, you're not talking about many atoms,' Fifield says. Archaeological sites seldom have the big, pure samples common on geological sites.

This sample, from Devil's Lair, a cave in Western Australia's south-west, was special. Fifield watched closely as the signals registered on computers in the facility's huge control room, sited 75 metres away from the accelerator to protect the scientists from radiation. The sample had undergone a new chemical pre-treatment method aimed at removing contaminating carbon and breaking the radiocarbon barrier. Michael Bird, then at the ANU, had developed the method, ABOX-SC (acid-base-oxidation, stepped combustion), and his colleague Chris Turney continued work on the technique, first at the ANU and then at the University of Wollongong.

They used it on charcoal from many levels of the Devil's Lair excavations, where previous work had run up against the radiocarbon barrier. Conventional processing in the 1970s yielded dates in the 30,000–35,000-year range. More recent AMS analyses of charcoal from the site using common chemical treatment delivered ages of about 45,000 years. The scientists got ABOX-SC radiocarbon ages of about 50,000 years on charcoal samples from below the deepest artefacts, and a similar OSL age on the sediments confirmed the result. They had crossed the event horizon. But the devil is in the detail.

Some scientists, including one of us, Gillespie, are not convinced that the ABOX-SC technique is as good as claimed, but its inventors stand by it. The dates of old samples carry big uncertainties because of contamination. A 50,000-year-old sample with just 1 per cent contamination from modern carbon would register an apparent age of 35,500 years. New laboratories, such as one at the University of Arizona, are being built especially for samples in the 40,000- to

60,000-year range. They have sample processing systems to minimise contamination.

The oldest mainland Australian archaeological sites include Riwi and Carpenter's Gap in the tropical Kimberley region of Western Australia, dated to 44,500 years and 44,000 years old respectively. Lake Menindee in semi-arid western New South Wales has a fireplace dated to 45,400 years, cross-checked with an OSL age on sediment heated by the fire. And Lake Mungo has burials OSL-dated to 40,000 years, while artefacts there and at Devil's Lair go back to almost 50,000 years.

Based only on calibrated radiocarbon dates from these archaeological sites, a 45,000- to 50,000-year date for the colonisation of Australia looks solid. Given that record, perhaps it's not too surprising that luminescence ages of 50,000 to 60,000 years from Arnhem Land unsettled some scholars. Archaeologists Jim Allen and Jim O'Connell have pushed the idea that the first Australians arrived not more than 45,000 years ago, dismissing some results from the unfamiliar new ABOX-SC and OSL techniques on archaeological grounds. This old idea of a short chronology fell on fertile, Eurocentric soil, on the assumption that modern humans must have arrived in Europe before venturing further afield. And with the debunking of the old dates for Jinmium, there was a view, arrived at through a curious form of reasoning, that if the correct date was not very old, it must be very young.

In the live-to-air broadcast of the Sydney Olympic Games 2000 opening ceremony, Australia was unable to put a single figure on the antiquity of Aboriginal culture. 'Over 40,000 years of culture, over 600 indigenous nations, over 200 Aboriginal groups, over 250,000 indigenous people, this is an awakening,' said Aboriginal actor Ernie Dingo in his commentary. Later, another commentator said: 'Despite their origins in ancient times, the Olympic Games are a youthful pursuit alongside the 60,000-year culture of our indigenous people.'

First landfall between 40,000 and 60,000 years ago raises the question of how the first Australians got here, and what challenges they faced in colonising the strange new land.

It was *terra nullius*, and you had to have a boat.

4 Stairway to heaven

On the north coast of Papua New Guinea an ancient staircase of coral reaches a kilometre into the sky. Each step is a remnant of a coral reef that once graced the floor of the Bismarck Sea.

Geologist John Chappell, of the Australian National University, and colleagues have been following the coral staircase on the Huon Peninsula for decades, in search of answers to questions about the first human colonisation of Australia. The answers hang on sea level, and when the migrants would or could have made the crossing from South-east Asia in the final leg of our species' migration from Africa to Australia. Those intrepid explorers would have taken one of two island-hopping routes proposed in 1947 by the American scholar Joe Birdsell—a northern route from Borneo to the 'bird's head' peninsula of West Papua, or a southern route via Timor to the north-west coast of what is now Australia. New Guinea, like Tasmania, was joined to Australia until about 14,000 years ago.

The terraces formed under water, with each step marking the death of a reef at a time of rapid sea-level rise, when water blocked the sunlight from algae in a symbiotic relationship with the coral. The microscopic plants harness sunlight to power photosynthesis, supplying most of the coral's food, and without them the coral dies. The terraces, which stretch about 100 kilometres along the coast, were exposed by the forces of plate tectonics. The north coast of New Guinea is at a subduction zone marking the boundary of the Australian continental plate and the West Pacific plate, and the terraces are slowly rising from the sea as the plates grind over one another. The rate of uplift is a few millimetres per year. At the end of an ice age, the sea level rises at a similar rate and the coral grows. At the end of the deglaciation the sea level is stable and the conditions of coral growth are similar to now.

The terraces represent glacial terminations that have occurred over hundreds of thousands of years. Broad, flat steps are interspersed with steep risers, documenting the height the ancient reefs reached as generations of the brilliant green, blue and red 'flower animals' settled on the skeletons of their predecessors. By dating the steps, Chappell could get snapshots of past sea levels. To evaluate sea level at the time of coral growth, he subtracted the amount of tectonic uplift from each step's present height above sea level.

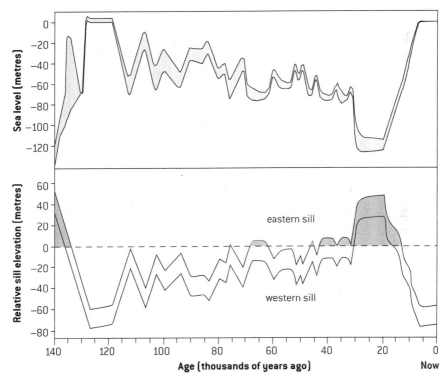

Sea level changes over the past 140,000 years (top panel), based on uranium-series dating of the Huon Peninsula coral terraces on the north coast of Papua New Guinea. Emergence and submergence of the Bassian landbridge between Tasmania and Australia (lower panel): shaded area represents land above sea level, showing that sustained pedestrian crossing was possible from 43 to 14 thousand years ago. Modified from Lambeck and Chappell (2001).

Chappell's interest in the formation was piqued by a photo in a geology journal from the 1950s, and he mounted an expedition to the site in the 1960s as a graduate student at the ANU. But when he reached the idyllic tropical islands off the peninsula, his hopes of getting to the steps, accessible at the time only by boat from the Vitiaz Strait, faded when wary locals refused to sell him a canoe. Things were looking up when the Melanesians invited him to a distant island for a festival punctuated by ceremonies in which immaculately costumed men acted out the role of sea spirits emerging from, and later retreating to, the waves. Still, several weeks of sensitive negotiations followed and Chappell's spirits sank as he whiled away the time. When the canny islanders agreed to sell him two dugout canoes, the asking price was beyond his meagre research funds, but the Melanesians finally admitted they were just testing him, and before long were waving the young geologist off on his journey. Chappell joined the two canoes with a platform, on which he mounted a sail, an outboard motor, an echo sounder and a winch. His tasks were to survey the modern reefs and lagoons, dredge up sediment samples, map the uplifted coral terraces, and collect samples from them for dating.

Back at the ANU in a bleak Canberra winter, he worked with Henry Polach on the radiocarbon dating of coral and *Tridacna gigas* (giant clam) shells from the youngest terraces. They found some of the results were compromised because the calcium carbonate mineral aragonite, originally precipitated by the living organisms, had partly changed to calcite, another calcium carbonate mineral with a different crystal structure. Recrystallisation had caused the radiocarbon dates to appear thousands of years too young. Chappell knew this because the ages were out of kilter with the coral terrace stratigraphy, and with uranium-series dates obtained for the same samples by his ANU collaborator Herb Veeh.

A naturally occurring radioactive isotope in seawater, uranium-234, is locked in the calcium carbonate skeletons of the coral polyps, where it decays to its daughter product, thorium-230, acting as an atomic clock. The clock works because of the different geochemical properties of these heavy elements—uranium is soluble in water, while thorium is almost insoluble and is not found in natural waters. This implies that all minerals precipitated from seawater are thorium-free, and all the thorium-230 atoms measured in the coral sample derive from the decay of uranium-234. The clock starts ticking at the beginning of the crystal's formation, when there are no thorium-230 atoms.

It takes 245,500 years for half of the uranium-234 atoms to decay to thorium-230, and 75,380 years for half of the thorium-230 atoms to decay to radium-226. After about 500,000 years the decay series reaches 'secular equilibrium', with the concentration of the parent uranium-234 isotope equal to the concentration of the daughter thorium-230 isotope. The clock practically stops. The maximum measurable age depends on the ability to measure with high precision the thorium-230 to uranium-234 ratio near this saturation point. Modern techniques such as thermal ionisation mass spectrometry (TIMS) work for ages up to 500,000 years. With the simpler technology available in the 1960s—counting alpha-particles, similar to the way Polach counted the beta-particles emitted by carbon-14—Veeh's limit was much lower, about 230,000 years.

Chappell compared the uranium-series ages with results on coral terraces from the West Indies island of Barbados, confirming that the chronology for major sea level changes was global. And the faster uplift rate on the New Guinea coast showed finer details of past high sea levels, which could be used to test theories about glaciation.

The Earth fluctuates between cold glacial periods, when water is locked up in the vast ice sheets that periodically shroud much of the northern hemisphere and Antarctica, and the warm interglacials, when sea level surges as the ice melts. The timing of these events puzzled geologists after the 1840s when Louis Agassiz identified their signature in the European and North American records. It was proposed that glaciations were linked to changes in the configuration of the Earth's orbit, but it was only in the 1930s that Serbian engineer and mathematician Milutin Milankovitch developed a full theory. He proposed that the cycle was driven by three variations in the Earth's orbit around the sun: eccentricity (the deviation from a circle), obliquity (the planet's axis of rotation relative to the orbital plane), and precession (a wobble in the rotation axis). According to Milankovitch's astronomical theory, these orbital variations have repeat periods of about 100,000 years, 41,000 years and 23,000 years respectively. In concert, they affect the amount and distribution of sunlight hitting the Earth's surface. It is now widely accepted that their impact is augmented by complex earth system dynamics. One feedback loop involves albedo—the amount of sunlight reflected from the Earth's surface, a parameter dependent on the extent of the ice sheets. Another relates to water temperature and involves the outgassing of carbon dioxide from the oceans. The early theories of the nineteenth century assumed that glaciations occurred when winters coincided with the aphelion (when the orbit of the Earth was at the farthest distance from the sun) and winters were longer or the solar radiation weaker. Milankovitch proposed that glaciations were caused by a reduction in solar radiation during summer in the northern hemisphere. This would preserve ice, which would accumulate in a growing ice sheet.

Milankovitch's model of glaciation was not widely accepted when Chappell started working the Huon terraces. It had begun to attract some support in the 1950s and 1960s when scientists studied cores from the Caribbean and Atlantic. The cores, cylinders of sediment drilled from the sea floor, contained the calcium carbonate exoskeletons of the single-celled marine organisms, foraminifera. There are two main kinds—'planktonic', in shallow waters, and 'benthic', in deep waters, each type comprising many species. Different species are adapted to different water temperatures, so the species distribution in the cores gives hints on climate. The primitive lifeforms also lock in climatological archives by preserving in the carbonate different isotopes of oxygen—a heavier one, oxygen-18, and the more common lighter one, oxygen-16.

In the early 1950s, American nuclear chemist Harold Urey saw a relationship between the oxygen isotope ratio in mollusc shells and sea surface temperature, and realised this could form the basis of a geological palaeothermometer. His Italian student Cesare Emiliani applied the method to foraminifera in deep sea sediments, and a series of papers beginning in 1955 showed a record of dramatic temperature changes in the past, stretching back 300,000 years. Emiliani introduced a numbering system for the ice age stratigraphy, called oxygen isotope stages (OIS). Odd-numbered stages denote low oxygen-18 warm periods, starting with the present interglacial, the Holocene (OIS 1). Even numbers refer to high oxygen-18 cold periods, starting with the Last Glacial Maximum (OIS 2). And between the extremes of glacials and interglacials are interstadials, periods with milder and relatively stable climatic conditions, which are also odd numbered.

Oxygen isotope stages offer a broadscale guide to global climate change. The most recent interstadial, OIS 3 at 60,000 to 30,000 years ago, embraces the time of first human colonisation and the annihilation of the megafauna in Australia.

Veeh and Chappell supported the astronomical model of glaciation in their 1970 *Science* paper on the Huon Peninsula coral terraces. Their record is still being refined as more and better dates come in. Milankovitch died in 1958, before seeing the Australian geologist's results and the full vindication of his orbital variation theory. A celebrated 1976 paper by James Hays, John Imbrie and Nicholas Shackleton, 'Variations in the earth's orbit: Pacemaker of the ice ages', published in *Science*, was the turning point. The scientists documented changes in the oxygen isotopes of forams, and the abundance of other microfauna, from deep sea cores located centrally between Africa, Australia and Antarctica. The measurements shed light on ice sheet volumes, sea surface temperatures and ocean circulation. By now the message was clear: Milankovitch's astronomical theory beat out a planetary rhythm for the Quaternary glacial cycles.

In the past decade, data from Greenland and Antarctic ice cores have run in counterpoint to the deep sea results. Cores more than 3 kilometres long display a frozen record stretching back more than 800,000 years. Continental-scale ice sheets form when seawater evaporates, falling as snow that is later compacted, and they are enriched in oxygen-16 because water molecules with the lighter isotope are the first to evaporate. The lightest isotope of hydrogen is also enriched in the ice. The isotopic ratios of oxygen and hydrogen march to the same orbital drum, their swings tracing past temperature variations. Despite its ice ages, the Quaternary saw glacial retreats as well as advances. Long periods of glaciation were punctuated by brief warmer interglacials, usually lasting less than 20,000 years. Between 2.6 and 1.1 million years ago, a full cycle of continental ice sheet advance and retreat took about 41,000 years, reflecting Milankovitch's obliquity signal. Since then, the cycle has shifted to a dominant 100,000-year eccentricity signal. The high-resolution deuterium profile from the EPICA Antarctic ice core confirms the interplay between obliquity and precession in the past 800,000 years, with the addition of a strong 23,000-year precession component for the last 400,000 years.

Although palaeotemperature records from the deep sea and ice cores are well established, land-based records are more difficult to pin down. Recently developed methods for studying glacial chronologies are based on long-lived cosmogenic isotopes such as beryllium-10 and aluminium-26, produced in surface rocks by secondary cosmic rays at rates of about 6 and 37 atoms per gram of quartz per year, respectively. Carbon-14 is also a cosmogenic isotope, turning up in rocks without first being filtered through the biosphere. Ultrasensitive AMS analysis of these rare isotopic tracers dates geophysical processes, including the waxing and waning of ice sheets, which shape the landscape over million-year timescales. Claudio Tuniz was one of the pioneers of the method in the early 1980s, when he used AMS to reconstruct the history of several geo-cosmic events, including the lunar and Martian origin of some Antarctic meteorites.

***

We inherit from Western Europe baggage about our planet's geography. Events are said to happen in the 'Middle East' or the 'Far East', and Americans live in the 'Western Hemisphere'. Western Europe was the centre of the known world, the 'Old World', birthplace of democracy and capitalism. East–west directions are measured from Greenwich, and that's where time starts, too. Australians live at the antipodes— about as far as you can get from Europe in space and time. Africa straddles the equator, and is joined by the Suez landbridge to Eurasia, while the now submerged Beringian landbridge from Siberia to Alaska straddles the Arctic Circle. During glacial times, Eurasia was joined to the Americas, which became united by the Panama landbridge just before the start of the Quaternary. Apart from brief warm interglacials, like the Holocene of today, Afro-Eurasia-America was one gigantic supercontinent. America finds itself in or out of supercontinental status as sea level is down or up, while Australia and Antarctica have been large island continents throughout the Quaternary.

Global-scale changes, for example in average temperature and sea level, force changes in biogeography at a local scale. Many large terrestrial animals, including humans, migrate long distances, on timescales from annual to those of Milankovitch's astronomical model, because climate dictates where they can make a living. And continental ice sheets periodically eliminate very large areas of land, only some of which is regained as sea level falls. The last Ice Age peaked about 21,000 years ago, when sea level fell to about 120 metres below today's level, and ended about 12,000 years ago. Chappell's updated coral record, with more precise uranium-series ages, documents rapid sea level changes in the OIS 3 interstadial—increases of 10 to 15 metres, with 'high stands' at about 30,000 years, 38,000 years, 44,000 years and 52,000 years ago. These surges, called Heinrich events, each taking only a few decades, were caused by the 'calving' of icebergs from northern hemisphere land-based ice sheets that saw huge volumes of fresh water dumped into the North Atlantic Ocean. The sea level jumps were small compared with the 120-metre change from glacial to interglacial, but big enough to drown islands and coastal plains.

Starting perhaps from the Ethiopian Rift Valley in east Africa, humans could have followed the rising sun and stayed in the tropics all the way to Australia.

Some say the sea crossings must have been made at times of low sea level, because the explorers were not smart enough to build boats equal to the long voyage, which would have involved a leg of perhaps 120 kilometres. The mariners would also have crossed Wallace's Line, the south-eastern boundary of the Oriental bioregion. The line, which runs between Borneo and Sulawesi, is named for Alfred Russel Wallace, who did fieldwork in the region while developing the theory of evolution he shared with Charles Darwin. Only birds, bats, rats, reptiles, shrews and *Homo sapiens* have made this journey, and they did it by flying, swimming, floating on storm debris—or by building boats. They would then have faced some big water barriers to cross Lydekker's Line, east of which, from New Guinea, is the Australian bioregion. And Chappell and ANU geophysicist Kurt Lambeck showed that crossing the Bassian landbridge from mainland Australia to Tasmania was possible only after about 43,000 years ago in OIS 3, so it is no surprise that Tasmanian archaeology has maximum dates just younger than that, while mainland sites go back to at least 50,000 years. New Guinea was also connected to Australia during ice ages, with Torres Strait flooded about 14,000 years ago. The oldest traces of people—stone tools in volcanic rocks deposited on a raised coral reef terrace on the Huon Peninsula—have been dated to between 52,000 and 61,000 years old.

Chappell and colleague Sue O'Connor, also of the ANU, argue that rising sea levels probably launched the voyages to Australia. Diverse, tropical coastal resources boom on coral reefs, in lagoons and mangrove forests, and on estuarine plains and wetlands when sea level is rising. Exploiting them would have produced maritime know-how. And as the rising sea drowned islands and coastal fringes, the exodus would to an extent have been forced. Perhaps the monsoon ferried coast-huggers in bamboo rafts to Sahul—also known as Greater Australia—the Pleistocene landmass comprising Australia, New Guinea and Tasmania.

It is unlikely that the site of landfall on Australia will ever be found. It is probably under water. The broad range of dates, centred near 50,000 years ago and accepted by most scientists as the time of first landfall, tallies with a time of rapid sea-level rise 52,000 years ago recorded in Chappell's coral staircase.

And, once they arrived, what did they do? Like colonisation debates, megafauna extinction disputes take esoteric turns, run up blind alleys, and stir passions political and scientific, raising heat and bulldust in keeping with a large portion of Australia's landscape today.

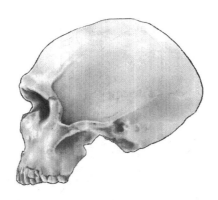

II
EXTINCTION

5 The melée

'If I'm a person trying to provide for my family, be it an indigenous Australian or a redneck in a ute [pick-up truck] in outback Queensland, I'm not going to spend a week trying to find a bilby if a two-tonne wombat is wandering past.' The speaker was a layman, his accent was broad 'Strine', and his comment was to a panel of experts fussing and fighting about who or what wiped out ancient Australia's giant animals, the megafauna. He must have thought he'd stumbled into a students' union meeting, complete with posturing, emotive rhetoric and personal abuse.

Most of the scholars at the public debate preferred to ignore the man's contribution; perhaps it smacked too much of down-to-earth Australian common sense. The topic at the forum, held at Canberra's National Museum of Australia in 2001, was a political hot potato, and had divided Australian researchers into two groups—those who blamed climate change and those who believed the animals had been killed off, in one way or another, by the ancestors of the Aborigines. The new cultural history museum was on a campaign to whip up interest in the Australia's Lost Kingdoms exhibition. Curated by the Australian Museum in Sydney and sponsored by the manufacturer of Yowie chocolate bars, the exhibition featured Australian megafauna, including a big bird billed as the 'Demon Duck of Doom'. Organisers had also signed Frank, one of Australia's most famous diprotodons, but his skull had been damaged in transit and they had to go with his understudy.

The NMA needed to do little more than ride the wave of publicity generated just a few weeks earlier by publication of research, led by Wollongong University's Richard Roberts, suggesting a human role in the mass extinction, which his team had dated to about 46,000 years ago. The date, soon after initial colonisation but at a time of relative climatic stability, suggested either a hunting 'blitzkrieg' by the first Aborigines, or a 'slow burn' when humans set the continent ablaze and the animals ran out of food. The team said climate could also have played a part.

The paper carried the authority of the prestigious US journal *Science*. The results had incensed academics in the climate catastrophe camp, who were running their own campaign in the mainstream media. In their sights was Tim Flannery, zoologist, then director of the South Australian Museum, co-author of the controversial *Science* paper, and one of the most deconstructed men in Australia. In his 1994 bestseller, *The Future Eaters*, he had argued a case for the rapid overkill hypothesis—blitzkrieg. The climate cabal had a big contingent at the NMA debate, which was to be broadcast on the ABC radio network's *Science Show*, hosted by journalist Robyn Williams, who had a tough time moderating the melée.

About 50 species, or 90 per cent of Australian land animals weighing over 45 kilograms, vanished late in the Quaternary. Most of the Australian megafauna were marsupials, but the giant flightless bird *Genyornis* and several large reptiles went, too.

Islands, great and small, also suffered extinctions in sync with the arrival of modern humans. The casualties include the half-tonne 'elephant bird', *Aepyornis maximus*, which vanished from Madagascar about 2,000 years ago, along with giant lemurs and other megafauna, and several species of the giant flightless moa, which disappeared from New Zealand only 600 years ago. An intact elephant bird egg washed up on a beach in Perth in the 1990s was dated by Tuniz to about 2,000 years old.

Extinctions hit other continents besides Australia. African animals came through relatively unscathed, perhaps because humans had evolved beside them, giving them time to lose their 'naivety' and gain hardwired defences against the top predator. Eurasia lost 36 per cent of its big animals, like mammoths, cave bears and hyenas, but the timing of extinctions seems out of sync with the arrival of modern humans there. And mammoths on Wrangel Island, north of Siberia in the Arctic, held on until 4,000 years ago. The debate over the Eurasian and African extinctions lacks the political charge of the extinctions debate in Australia and America, where indigenous people are still in decolonisation struggles against a culture indivisible from science, and where some project on indigenous cultures the image of Pleistocene Greens. Neanderthals vanished on the arrival of modern humans in Europe, but categorising the European Neanderthals as megafauna raises the ire of some.

The Americas, hit hard by the catastrophe, lost their giant ground sloths, sabre-toothed cats, lions, cheetahs, mastodons, mammoths, camels and several species of horse. American experts still argue over whether the ancestors of Native Americans had a role in the extinctions or it was all down to climate. There has been leeway for the protagonists because the first unarguable settlement dates for the Americas coincide with a time of rapid climate shift. But first settlement in Australia was not at a time of dramatic climate change—raising the prospect of disentangling climatic impacts from human ones. Work on the ancient DNA of one American species underlines the pitfalls of dismissing climatic impacts, but the Australian story is clearer.

***

Something out of the ordinary had happened to the diprotodons, the meat-eating 'marsupial lion' *Thylacoleo carnifex* and several genera of giant kangaroo. The debate about what did them in has been raging since the early nineteenth century, when naturalists first described the strange fossils being unearthed by colonial explorers in the antipodes, like Thomas Mitchell, Surveyor-General of New South Wales between 1828 and 1855. Many of the fossils were found in caves that had acted as pitfall traps. Others were on plains and lakebeds, including Lake Callabonna in South Australia, where an Aboriginal stockman discovered a treasure trove of fossils in the early 1890s.[1] One of the most active researchers of the period, the British anatomist Richard Owen, who sank decades of his life into the study of the fossils, said he suspected 'the hostile agency of man' in the extinction.

New finds were being made into the twentieth century, and there are now dozens of major sites spread across Australia and New Guinea. Climate catastrophists blame the last Ice Age for the beasts' demise, but have yet to explain why the megafauna, veterans of some 20 earlier ice ages, yielded only to the most recent one.

Those arguing for a human role are divided between the blitzkriegers and sitzkriegers, who propose indirect human impact. The blitzkrieg hypothesis was formulated in the 1960s by palaeoecologist Paul Martin, of the University of Arizona, to explain the demise of the American megafauna. Tim Flannery, the most public Australian proponent, says people wiped out the big animals, which, he says, had a major ecological role in recycling nutrients into the soil. Fire-promoting plants that can tolerate infertile soils gained ascendancy. Wildfires broke out and the hydrology changed, altering the climate. Flannery says that the ancient Aborigines developed land management practices in response, using small fires to prevent larger ones. In an impoverished ecosystem, the people reached a new balance that would only be upset again by the arrival of Europeans.

To the sitzkriegers, humans lit fires to clear pathways or flush out prey, sparking a mass extinction. Rhys Jones, who coined the term 'firestick farming', and palaeontologist Duncan Merrilees developed the idea about the same time in the late 1960s. 'We must not be dazzled by the power of our own technology to underestimate or ignore that of the most primitive one,' Jones wrote in 1968.

In the same year, Merrilees published his views in a now famous paper, titled in those more candid times, 'Man the destroyer: late Quaternary changes in the Australian marsupial fauna': 'If he [Aboriginal man] used fire indiscriminately and expected that its repercussions would not be excessive, he may have been much mistaken, and may have been punished in that most decisive way, by diminution of his food supply and decline in his own numbers …'

Merrilees now says his paper 'stirred the possum', but stresses that more recent evidence shows that Aborigines were much more discriminating and careful in their use of fire than was obvious from accounts in the literature at the time he wrote the paper. 'It sounded to me as though they were gay and carefree and just lit up whenever they felt like it, but that's not so. That's not to say that the progenitors of Aborigines when they got into Australia weren't carefree and indiscriminate. It isn't justifiable to assume that they were. That's what I think now.'[2]

Without secure dates on megafaunal remains and on human colonisation, as well as a handle on the palaeoclimatology, all three models—climate change, blitzkrieg and sitzkrieg—remained pure conjecture. One of the first attempts to date megafauna bones was in the mid-1950s, when American palaeontologist Ernest Lundelius, then at the University of Chicago, deployed the newly developed radiocarbon technique on the problem. Lundelius, a leading proponent of the climatic causes of extinction, dated megafauna bones from Mammoth Cave in the Margaret River wine-growing region of Western Australia. The site is a census of the late Pleistocene fauna of the state's southwest, recording everything from diprotodons through big kangaroos and echidnas to *Thylacoleo*. He recovered a small sample of charcoal from near the bones and persuaded the exploration department of the Humble Oil and Refining Company in Houston, now Exxon-Mobil, to date the charcoal and thereby indirectly date the bones. Oil companies wanting to find out how oil reserves formed had been quick to set up radiocarbon dating labs after the development of the technique.

The date turned out to be more than 41,000 years, beyond the reliable limit of the method at that time.[3]

Several other sites were dated over the next four decades, with some yielding ages young enough to suggest a long overlap of humans and megafauna, apparently clearing people as the culprits in the mass extinction. The results were suspect, however, partly because of the radiocarbon barrier and problems with the indirect dating of fossils, all of which had failed to yield any collagen for direct dating. And nobody really knew when humans had first colonised Australia. If researchers could not accurately date either the arrival of people, or bones older than 40,000 years, they could not settle the argument. As in America, the knowledge gap created a vacuum, to be filled by politics.

'... Australian indigenous peoples today ... are among the most noble, most responsible managers of sustainable harvesting programs in Australia,' Mike Archer, then director of the Australian Museum, told the 2001 NMA forum. Archer is a highly respected palaeontologist known for his work on ancient marsupials and for a project he launched to clone the thylacine—the famed Tasmanian 'tiger'. He is a former teacher of blitzkrieger Tim Flannery, but now his nemesis.

'For at least 36,000 years, they kept alive and well species which we managed to exterminate incredibly rapidly in just the last 200 years,' Archer added. 'You'd have to ask why their ancestors have had such a totally different ethic and allowed this massive extermination of animals across the whole continent.'

Robyn Williams: 'Because they hadn't learned yet, possibly.'

Archer: 'There's an argument about that, but I think it may be unnecessarily judgemental. You'd really want to see good evidence before you assumed that you could accuse them of this dreadful bit of vandalism ... I'm focused on what did they do that kept the biota alive that we're promptly destroying ... I think there is more to learn from the survival of Australian animals and what enabled that to happen.'[4]

The extinction question is mired in the politics of land management today, especially the management of Australia's national parks. Many have been returned to regular control burning regimes, similar to those once used extensively by Aborigines, to protect them from wildfires.

'A large number of the Australian public are concerned about, and have guilt feelings about, the extinction of animals in the last 200 odd years, and it calms some people's minds to have the feeling that ... if Aborigines were doing the same thing, it's all right,' the ANU's Alan Thorne told the forum. 'I've already heard a number of people argue that, "well, these people were wiping out animals in the past, why would we let them have national parks back?"'

The voices of Aborigines have been conspicuously absent from this 'discourse'. Perhaps some consider the image projected on them a bit ordinary, especially when it portrays their ancestors as scavengers and incompetents, who walked only lightly on the landscape.

'David Bowman, this fascinating character working in the Northern Territory, was challenged with the job of exterminating water buffalo from Kakadu National Park,' Archer told the forum. 'He had helicopter gunships at his disposal. He had every bit of vicious weaponry known to mankind available to do this simple task of exterminating huge animals like water buffaloes from one park. Eventually he gave up. He couldn't do it. And at an extinction conference a couple of years ago, he basically challenged us to think how then could indigenous people who came into Australia a long time ago armed with stone tools, maybe spears, how could they have exterminated 46 species, untold millions of individuals, across the whole continent and New Guinea and all the islands around Australia. And he frankly couldn't understand how we could have possibly conceived that it could have happened.'

In fact, the Kakadu 'buffs' share tens of thousands of years of evolution with humans in Asia, and it takes more than a couple of years for a blitzkrieg. And Aborigines—the best trackers in the world—will tell you there's more than one way to skin a *Thylacoleo*. Bowman's own computer modelling research conducted a few years later would wind up supporting a blitzkrieg.

Meanwhile, the Darwinian struggle between academics who share a lot of history makes the debate even hotter, and dimmer. Some researchers hit back when their results are challenged. The formalised sparring of science, with protagonists speaking candidly and presenting their arguments in peer reviewed journals or at scientific meetings gives way to public stoushes in the mainstream media. Reasoning gives way to polemics, key results are played down, straw men are set up to be knocked down. And researchers are the targets of *ad hominem* assaults on their personal and scientific reputations. Such behaviour, once unacceptable in academia, is legitimised in a political world view. One reviewer of a proposal for a project on megafaunal extinctions wanted the research proponents to rule out a blitzkrieg before gathering their data, and to engage Aboriginal leaders in the interpretation of the results. The reviewer wrote:

> White science which ... has already blamed indigenous arrivals with disposing of the megafauna runs a risk of perpetuating a sad tradition in which studies involving indigenous cultures have been advanced without clear understanding from the indigenous people affected by them.

There has already been a significant backlash against science that has advanced the view that Aboriginal overkill was responsible for mass extinctions ... There is clearly a case here where scientific objectivity needs, not only to be a foundation, but clearly seen to be such foundation. It might have been preferable if this proposal had distanced itself from some previous claims which have lent fire to these embers sensing past imperial injustices.

Only the strangest, fuzziest thinking holds that 'scientific objectivity' is best served by ignoring the data and framing a hypothesis on the basis of current Aboriginal disadvantage. The great megafauna extinction debate is beset by would-be critical theorists who look at the ancient world and see only modern power struggles. The scientific juggernaut rolls on, however. Some researchers have already moved from squabbles over human agency to attempt to discriminate between a blitzkrieg and a sitzkrieg. Some are seeing if El Niño compounded human impacts. A dating revolution has advanced the debate over the past few years, and bones, eggshell, pollen, charcoal, dung, ancient DNA and silicon chips have provided other vital clues.

In his seventies, and 40 years after obtaining his original radiocarbon date of more than 41,000 years for megafauna, Ernest Lundelius once again entered the Margaret River caves. This time, in the late 1990s, he was an observer on an expedition with the Roberts team. The team was armed with a new dating technique, a strategy to tackle problems in indirect dating, and a big-picture approach to prehistory that was gradually replacing investigations fixed on individual sites. The work would take in sites across Australia and New Guinea. Roberts and colleagues would distil their findings into the 2001 *Science* paper that caused all the fuss at the NMA.

However, some of the answers were already in. They had come not from the remains of the charismatic giant marsupials represented in the Mammoth Cave deposit, but from the eggshell of the extinct giant bird, *Genyornis*, in a most amazing story of the role of serendipity in scientific discovery.

6 Inside Geny's eggshell

Fifty-five thousand years ago, a *Genyornis newtoni* chick broke through the eggshell that had confined it for two months and stood up on a sand dune on the edge of South Australia's Lake Eyre.

The genes it carried suited it exquisitely to an environment slowly drying out. The chick was a member of the last of eight species of Dromornithidae, a family of Australian birds related to ducks and geese that had traded the ability to fly for large body size. This evolutionary transaction bought them real estate in the Centre, where animals that could process lots of poorer food had an edge.

The ancestral dromornithid was a flying bird dating from the Eocene, 50 million years ago, and probably inhabited the coast. The Lake Eyre chick, whose species is thought to have made its appearance in the Pleistocene, inherited its ancestor's nasal salt glands, an adaptation that enables seabirds to drink salt water and that would have conferred an advantage in the increasingly saline environment of the lake.

When it grew up, remarkably quickly, the big-headed Geny would have stood 2 metres tall on sturdy legs and weighed 200 kilograms. It would have ranged across savannah grasslands and shrublands in search of grass, grasshoppers and leaves. Its world was poor compared with the extravagant forests on parts of the coast at the time, but lush in comparison to the desert landscape of Lake Eyre today. The lake, with a catchment covering a sixth of the continent, had been up to 25 metres deep but was shallower in Geny's time. It was still able to support colonies of the birds, however, and the sand dunes around it harboured their silky, cream, 15-centimetre-diameter, almost spherical eggs.

The red-billed chick flapped its vestigial wings as a fresh wind blowing off the water ruffled its feathers, and it looked out through beady eyes on a weird land of giants.

At 1.5 metres tall and more than 3 metres long, *Diprotodon optatum* was the biggest marsupial ever. These lumbering, furry creatures outnumbered their close but smaller relatives, *Zygomaturus trilobus* around Lake Eyre. Short-faced kangaroos boomed across the country. The biggest, the 2.5-metre-tall *Procoptodon*, which at more than 200 kilograms was three times as heavy as a modern 'big red' kangaroo, shared the habitat. *Phascolonus gigas*, a bulky burrowing wombat weighing at least 150 kilograms, excavated the land around the lake. A jumbo Tasmanian devil and the ferocious marsupial lion, *Thylacoleo carnifex*, stalked prey. *Thylacoleo* was as big as a leopard but was a mere pussycat compared with giant reptiles that had beaten the mammals to the top of the food chain. The 7-metre, 1-tonne goanna, *Megalania prisca*, made short work of any herbivore that got past *Wonambi*, a 30-centimetre-diameter snake stretching 6 metres long.

Living in the shadows of the giants were the ancestors of animals alive in Australia today. The *Genyornis* nesting grounds were next to those of ancient emus, which resembled the bigger birds but were not closely related. The landscape was littered with their pimpled green eggshell. Water birds made a din. Ducks, pelicans and a now-extinct species of flamingo flocked to the vast lake and the ancient rivers that fed it. Tortoises and crocodiles broke the surface of the water.

But this strange Pleistocene world was about to be shattered.[1]

***

Late in the Holocene, geologists John Magee and Gifford Miller head to Lake Eyre, now a 10,000-square-kilometre desert saltflat most of the time. The last stop is the Oasis Café in Marree, a 700-kilometre drive north from Adelaide. This tiny outback town is where the old Queensland–South Australia stock route, the Birdsville Track, meets the Oodnadatta Track. It is now more famous for the Marree Man geoglyph, a 4-kilometre-long figure carved into the ground in the late 1990s by pranksters using earthmoving equipment and GPS.

Magee checks the shed housing his drilling gear. It is known as 'the ANU Marree campus', as if competing in the irony stakes with the bit of scrub opposite signposted 'MCG' (for Melbourne Cricket Ground). On the road out of town, a huge sign warning 'Remote Areas Ahead' spells out the perils awaiting the traveller uninitiated in this land. The place gets less than 125 millimetres of rain a year, the lowest reading in Australia, and summer temperatures soar over 45 degrees Celsius. The sheep and cattle stations in these parts run on water from the Great Artesian Basin.

The road to Lake Eyre cuts through land able to carry only a thin cover of wattle and saltbush. Even these hardiest of plants are stunted. A gate in the dog fence opens the way to the dingo side of the 2,200-kilometre-long barrier erected to protect sheep on properties in the state's south. It is the gateway to the wasteland. Shingleback lizards, a few kangaroos, a black kite and red-capped plover are the only wildlife in sight. Magee and Miller, of the University of Colorado at Boulder, drive past Lake Eyre South, joined by the Goyder Channel to the incomprehensibly vast northern section of the lake. They pull up on the southern margin of Lake Eyre North at Williams Point, named for a young scientist who died before completing his research on the lake.

The scene before them is more like a watercolour than the picture painted in the South Australian Government's tourism promotional material of a life-threatening, 'stark, inhospitable wilderness' that evokes fear and a sense of personal insignificance. (The geologists have never sighted a tourist in years of expeditions to the area.) Soft clouds break a sky that gradates from a pale blue to a blue wash. In the foreground, the brilliant white of the salt is streaked by the soft gold of the sand. But take your sunglasses off and this subtle scene hurts your eyes. Technicians drilling sediment cores from the centre of the lakebed, where the salt reaches half a metre deep, wear welding goggles.

This is a place where the light plays tricks on the eyes, where mirages erase the horizon. To the north-west, Shelly Island, which looks stranded, seems like a short stroll from the shoreline. But it is 6 kilometres away, and as you crunch over the salt crust towards it the island remains tantalisingly out of reach. John Magee's benchmark, a steel star picket driven soundly into the bank, is a solid frame of reference in this ambiguous place. Another anchor is 20,000 kilometres overhead, where satellites beam data down to the geologists' GPS receiver.

Behind the point, Magee and Miller walk in straight lines across an ancient sand dune, into the sun and against a wind that feels like a sea breeze. An emu recently laid an egg nearby, but the big birds are elusive. *Genyornis* eggshell is scattered over dry sand mined by rabbits: the geologists have recorded the dune's GPS coordinates. A piece of the shell in which the Geny chick once started its life glows in the sunlight, and the two put it into a numbered plastic bag. The fragment, and thousands like it, will shed light on the extinction of the chick's species.

***

Neither John Magee nor Giff Miller remembers when they plotted the histogram showing the frequency of emu and *Genyornis* eggshell samples against the ages of the fragments. The geologists had been using the samples to date sediments from Lake Eyre for research on the Australian monsoon. They were investigating the swings in rain dumped on the interior by summer winds over the past 120,000 years. Monsoonal rains fall only lightly on Lake Eyre, surrounded now by the Tirari and Simpson deserts, but they once charged mighty rivers—like the Diamantina and Cooper—in its vast catchment to the north-east, deep into Queensland. Nowadays, the monsoon usually only touches the northernmost parts of Australia, and Lake Eyre stays dry most of the time.

The rainfall records were written in the advance and retreat of the shoreline of the lake. Other archives were in the waxing and waning of dunes that grew and eroded around the lake as the wind hurled sand from the lakebed in dry phases. In extreme arid periods, the ground water, which rises to the lakebed surface through tiny channels between the sediments, becomes saltier. Salt crystallises in the sediments, loosening them up and exposing them to wind erosion. The lakebed, which is now 15 metres below sea level and the lowest point on the continent, drops by an amount that betrays the severity of the dry.

Tracing the effect of water, wind and sun on the lake is like analysing the harmony of a baroque fugue, however, and Miller and Magee read Lake Eyre's complex hydrology like maestros reading a Bach score. Timing is everything—in music and in serious prehistory—so the pair needed a reliable way to date the sediments. To find out when a layer was deposited, they dated eggshell from that layer.

Since beginning their collaboration in 1992, Miller and Magee have collected about 100,000 eggshell samples from around Lake Eyre and other sites. Their field trips have been in winter, when the top temperature is a mild 25 degrees Celsius. The two, backed up by other team members, would spend up to six weeks in the desert, driving to sites each day and combing the dunes and lunettes on foot or on motortrikes for the precious eggshell scatters.

The eggshell of big flightless birds, which lay huge, strong eggs, resists chemical attack. It persists in the environment for millions of years, as long as it is buried and protected from wind abrasion and animals wanting a quick calcium fix.

The scientists scanned aerial photos and satellite images to find prospective sites. What they were looking for were 'blowouts'—places where the wind had stripped sand from the dunes, exposing heavier objects. Blowouts in progress often uncover samples in situ, and occasionally entire nests emerge. Going by the weight of the fragments, the nests, often in groups, contained up to 15 eggs, suggesting that the birds hatched their eggs communally or returned to the same nest each season.

As Miller and Magee worked the biggest deposits, they recorded each site's stratigraphic details and GPS coordinates. Sometimes they took sand samples for optically stimulated luminescence dating. The geologists noticed that the *Genyornis* eggshell was usually near the emu's. They were growing curious about the extinct bird that occupied the same dunes as the ratite, but for now their attention was fixed on the sediments, not the means of dating them.

Miller's laboratory in Boulder dated the samples through a technique known broadly as amino acid racemisation (AAR). The calcium carbonate of eggshell is formed on a scaffold of protein, which, along with their shape, lends eggs their strength. Isoleucine is one of the amino acids in the protein. Since the isoleucine molecule morphs at a known rate into a molecule with a different structure, D-alloisoleucine, the protein works as a chemical clock. Miller's lab used a sensitive analytical technique—high pressure liquid chromatography—to separate the alternative forms, or 'enantiomers', of the amino acid and measure their proportions in each sample. The lab ran 24 hours a day to get through the work. AAR is relatively imprecise, so the geologists processed huge numbers of samples in the way that epidemiologists do metadata analysis to correct for confounding factors in statistics on modern diseases. The analytical technique is quick and cheap compared with other methods, such as radiocarbon dating, which costs more than $1,000 a sample. The team would wind up processing more than 2,000 samples from Lake Eyre alone.

Since the rate of isoleucine's structural change depends on ambient temperature, the chemical clock runs faster at some sites than at others. It also varies over the millennia in time with global climate change: temperatures in the Centre were up to 9 degrees Celsius lower during glacial phases of the Pleistocene than now. Magee and Miller had to calibrate the clock with a series of radiocarbon and uranium-series dates on eggshell and optically stimulated luminescence dates on sand from which the shell was recovered.

Then, the two plotted all their eggshell data, Geny and emu. 'Suddenly, there it was,' recalls Magee. 'There was a continuous distribution of emu to the present, but *Genyornis* just went bang and stopped at 50,000 years. We realised that inadvertently we'd dated the extinction of *Genyornis*.'

The unbroken record of emu eggshell ruled out the possibility that a preservation problem was behind the abrupt disappearance of the bigger bird's eggshell, so the two got to work firming up their data. They included two semi-arid regions—Lake Frome to the south-east of Lake Eyre and the Murray–Darling Basin of western New South Wales—to see if the extinction was widespread or only local. The

Murray–Darling region has one of the most dependable water supplies in the interior, and would have been a refuge for megafauna during droughts.

The combined results for the dataset, which stretched back 130,000 years, dated the extinction to between 45,000 and 55,000 years ago across all climate zones. The number was close to the date of colonisation of the continent. But to find out whether humans or climate change had triggered the extinction, the two had to check the palaeoclimatological records. They drew on their own work on the monsoon at Lake Eyre. The records show that 60,000 years ago was the last time before the mass extinction that the lake was a permanent water body, so it was drying out gradually when *Genyornis* vanished.

However, the biota supported by the lake had been through tougher times, the hardest being 140,000 years ago when the wind scoured sediments from the lakebed, wearing it down to 19 metres below sea level, lower than during the last Ice Age and now. Other, less intense, dry periods were between 60,000 and 50,000 years ago and between 30,000 and 16,000 years ago. *Genyornis* vanished from the Lake Eyre catchment during what was either only a moderately dry period or a subsequent wet period.

Work led by Jim Bowler on the Willandra Lakes pointed to a prolonged dry in the neighbouring Murray–Darling catchment. That dry broke 60,000 years ago, with rains filling the basin with permanent water and vegetation flourishing. About 40,000 years ago the climate abruptly became drier again, and then fluctuated before the lakes completely dried out about 25,000 years ago. *Genyornis* was extinct in the Murray–Darling region long before the onset of aridity 40,000 years ago.

The palaeoclimatologists had unearthed the first hard evidence that eliminated climate change as the cause of the mass extinction of the megafauna. After devoting years of their lives to the study of the monsoon, they were about to broaden their research. Magee and Miller announced preliminary results at the Quaternary Extinction Symposium in Perth in 1997 and published their final data in a seminal paper in *Science* in 1999, breaking a 20-year drought on the publication of megafauna extinction research in major journals. It took a lot of hard science, a dating revolution and a bit of luck to get the subject back into the mainstream.

In a repeat performance, the two caused a stir at the next meeting of extinction buffs, in 2005, with the announcement of further results, duly published in *Science*. The *Genyornis* eggshell had shone a torch on the past, this time illuminating the ancient landscape. It revealed a scene of fire and devastation as modern humans spread across the continent.

The scientists had extended their research on Lake Eyre and the Murray–Darling Basin to include samples from as far afield as Port Augusta, South Australia, where a site dubbed 'Geny Heaven' is thick with eggshell. They wanted to confirm that the Lake Eyre results were reflected outside the semi-arid zone. They analysed more samples from each region and looked back further in time, to 140,000 years ago. The work supported the 50,000-year extinction date across the different climatic zones. Now convinced that humans must have had a lot to do with the extinction, the scientists wanted to get a fix on the mechanism—a blitzkrieg or environmental destruction. Magee and Miller needed to find out what *Genyornis* ate and how it made its living. In another case of serendipity, they found a way—a technique that had been used to reconstruct the diets of everything from elephants to Vikings.

Radiocarbon labs analyse samples for the stable isotopes carbon-13 and carbon-12, along with the carbon-14 used for dating. Most organisms discriminate against heavy isotopes like carbon-13 and carbon-14 in a process called fractionation, the natural equivalent of the enrichment of uranium used in the production of nuclear weapons. Fractionation can throw radiocarbon dates out, so laboratories make a correction, using carbon-13 as a guide to the strength of the organism's discrimination against the radioactive isotope. In their earlier work, Miller and Magee had noticed high carbon-13 levels in samples with ages nudging the radiocarbon barrier. 'There was a clear signal,' says Miller. They enlisted Marilyn Fogel, a biogeochemist at the Carnegie Institution of Washington, to investigate the possibility of using carbon-13 and carbon-12 to reconstruct the feeding strategy and environment of *Genyornis*.

Since plants favour the lighter isotope, the sugars they produce have a lower carbon-13 to carbon-12 ratio than atmospheric carbon dioxide. However, they differ in the extent to which they fractionate, according to their mechanism of photosynthesis. The carbon isotopes pass through the food chain. Their proportion in an animal's eggshell, teeth or bones reveals what vegies were on the menu—trees, shrubs and some grasses, like wheat, which use the so-called 'C3' photosynthetic mechanism, or C4 plants—drought-adapted grasses that conserve water, including corn, sugarcane and Australia's kangaroo grass. Favoured by hot climates and high levels of solar radiation, C4 grasses predominate in the Top End, accounting for 95 per cent of the cover there. They also have an edge on C3 plants in arid climates, as the success of spinifex, an unpalatable desert C4 grass, attests. And they can handle low levels of atmospheric carbon dioxide.

The C3 mechanism arose in the first photosynthetic micro-organisms perhaps as long as 3 billion years ago, while the C4 route evolved several times after 25 million years ago when the Earth's carbon dioxide levels plummeted, says plant biochemist Hal Hatch. Hatch and colleague Roger Slack were among Australians who did pioneering work on C4 photosynthesis in the late 1960s and early 1970s. Working for the Colonial Sugar Refining Company, the two were, with Russian and American teams, co-discoverers of the deviant process. They were the first to recognise the significance of C4 photosynthesis and to nail its unusual mechanism.

Graham Farquhar, another Australian photosynthesis expert, renowned for esoteric work modelling the reactions, states, inaccurately, 'This is really easy!' before launching into an exposition of the deep chemistry of isotope fractionation. In his office at the ANU, he leaves out the mathematics that baffles even his colleagues as he explains the formidably complex processes in the chaotic green world inside the leaf. They include the enzymes catalysing photosynthesis, and the differing thermodynamic properties conferred on carbon dioxide by carbon-12 and its lumbering carbon-13 counterpart as the gas enters and exits the leaf stomata, or pores.

Fractionation starts when plants grab carbon dioxide molecules from the air in the first steps of the photosynthesis 'dark reactions'—

the food production line powered by solar energy harnessed in the 'light reactions'.

C3 plants use the huge, 45,000-atom enzyme rubisco to catalyse these initial steps. The protein is packed in the chloroplasts, site of food production in the 'mesophyll' cells inside the leaf. Rubisco cannot distinguish well between carbon dioxide and oxygen, however. When the plant partially closes its stomata during scorchers to stem water loss, the carbon dioxide in the leaf's air pockets is depleted. Rubisco reacts more with oxygen and less with carbon dioxide, and photosynthesis and food production are stifled.

The C4 grasses have an evolutionary advantage over their C3 counterparts, conferred by the enzyme used in the first step of carbon fixation. They use rubisco, too, but take a different chemical pathway in carbon fixation, using the enzyme PEP carboxylase to front the first steps. This enzyme, also in the mesophyll cells, does not react with oxygen. An acid product of it delivers carbon dioxide to rubisco, which in C4 plants is in different specialised cells, overwhelming it with the gas and compensating for its lack of discriminatory power. C4 plants can partially close their stomata on hot days without damping photosynthesis. 'C4 has a turbocharger up front,' says Farquhar.

C3 plants discriminate more strongly than C4 plants against the heavier carbon isotope. To measure the difference, Marilyn Fogel and colleague Beverley Johnson collected plant samples from around Australia. They had to get a handle on some of Australia's unique grasses, which number in the hundreds. The scientists also conducted feeding experiments on modern emu, ostrich and quail to see how the isotopic signature of plants was expressed in eggshell. Carbon in the calcite minerals in shells comes from dissolved bicarbonate in the blood, and reveals the bird's diet in the days to weeks before egg laying. Carbon in the eggshell protein scaffolding comes from stored protein, and records diet in the months before laying.

Fogel can read isotopic signatures like a menu. She analyses finger-nail samples from scientists visiting her lab, adding the results to her database. Like most Australians, Magee displayed carnivorous tendencies, while Miller carried a stronger C4 signal, reflecting Americans' penchant for corn.

Team member Michael Gagan, also of the ANU, measured the isotopic ratios on a mass spectrometer of which he is visibly proud. He analysed samples, each weighing just 200 micrograms, which were punched out of the eggshell, along with other tiny discs destined for AAR and, sometimes, radiocarbon analysis.

In the instrument's high vacuum, which is sealed tightly against the atmosphere, two drops of phosphoric acid liberated carbon dioxide from the sample. The gas was isolated and streamed into a chamber where a hot filament emitted electrons that bombarded the gas, ionising it. Accelerated past a powerful magnet, the positive ions changed course, their pathways dictated by their masses. Two beams, one for carbon-12 and one for carbon-13, hit collectors, which counted the amounts of each isotope.

As sensitive as they are, mass spectrometers are not stable enough from day to day to give absolute numbers, but measure isotopic ratios relative to a calcium carbonate sample set as an international standard. The first standard was Peedee belemnite (PDB), a fossil from the east coast of the United States of a now extinct squid-like creature. That source of pure, uniform calcium carbonate has long been exhausted and scientists now use a different one, called NBS19, said to be a marble toilet seat of obscure origins. They use it sparingly. 'We don't want to run out of NBS19,' says Gagan. 'It's a very limited toilet seat.'

From the numbers, Fogel could see that before 50,000 years ago, emus ate C3 or C4 vegetation or both, and had the flexibility to switch between nutritious C4 grasses and C3 trees and shrubs. But their *Genyornis* contemporaries needed at least some of the drought-proof C4 plants to survive. For the past 45,000 years, C3 plants have formed the emu's staple food, suggesting an abrupt and widespread destruction of the C4 vegetation, or at least the palatable components of it. One C4 plant, canegrass, flourished after the ecosystem collapse, but most animals give it a miss.

The team had long suspected that ecosystem collapse was behind the *Genyornis* extinction. If so, all herbivores should be affected, and wombat teeth were the best prospect for testing the prediction. The scientists analysed wombat tooth enamel from specimens from Port

Augusta and the Murray–Darling basin to determine the grasses and reeds that ancient marsupials were eating. The results were the same as for *Genyornis*. The drought-adapted C4 plants accounted for 40 to 100 per cent of the diet of wombats, obligate grazers, living 50,000 years ago; but 5,000 years later, wombats were eating mainly C3 vegetation.

The research was bolstered by work by Rebecca Fraser, previously of the ANU, who compared the stable isotope signature in the teeth of wombat roadkill from Queensland, where C4 grasses predominate, and the C3 landscape of the Snowy Mountains in southern New South Wales. She found that the isotope signature strongly differentiated between the food sources.

In more temperate zones, studies of ancient pollen can reveal which plant species covered the landscape, but little pollen is preserved in sediments of the arid Australian interior. The *Genyornis* and emu eggshell analysis filled the gap, and put the first hard date on the creation of Australia's dead centre. With the matching work on wombat tooth enamel, the scientists had identified a massive ecological crash, and the crash coincided with the arrival of humans. To see if climate had also played a part, the team split its dataset into 15,000-year intervals stretching back 140,000 years. This sharpened the focus on times of Ice Age climate change, from the previous to the present interglacial.

Climate had not sparked the collapse, however, as the drought-resistant plants had withstood past climatic fluctuations. It took the shock of human colonisation to kill them off. They never recovered, not even in the wetter, warmer conditions at the start of the current interglacial, 12,000 years ago.

Uncertain dates for the mass extinction, from 45,000 to 55,000 years ago, and for the entry of people, also 45,000 to 55,000 years ago, prevented the scientists from discriminating finally between the two possible pathways to extinction. Perhaps it was the early Australians' use of fire that delivered the fatal blow. The megafauna might have held on for 10,000 years after people landed—a 'slow burn' to extinction. Still, no-one knows how long it takes to destroy an ecosystem with fire. Miller and Magee have recovered fewer than 10 fragments of burnt *Genyornis* eggshell that look like they could have been burnt on hearths. They have found many fragments of emu eggshell leftovers, however, suggesting a short overlap between people and the bigger bird.

And perhaps big animals expired soon after people made landfall, in Flannery's blitzkrieg, with fires raging in the aftermath: the human figures in the view of the past lit up by the *Genyornis* eggshell were blurred. Were the ancient people holding firesticks or hunting spears?

Miller and Magee favour the slow burn scenario, saying overhunting would not cause 'the dramatic changes at the base of the foodweb' revealed in their datasets. A blitzkrieg would not cause the dietary shift seen in emus and wombats. And there is no strong empirical evidence from islands, where rapid overkill is generally accepted as the cause of extinctions, that widespread vegetation change inevitably follows a blitzkrieg, they add. Computer modelling by the team, reported in the journal *Geology* in 2005, suggested that human-lit fires could alter the climate on a continental scale. Their theory is that the destruction of the vegetation would have reduced the amount of water returned to the atmosphere by plants and delivered to the interior by the summer monsoon—the original focus of their research. The monsoon carried water to Australia's heart before 60,000 years ago, but now usually drops rain only along the north coast as dry winds beat the Centre. Similar rainfall systems on other continents came and went with the ice ages and were recharged by 10,000 years ago, but the Australian system stopped dead in its tracks. The crash in plant diversity wiped out the specialised herbivores, such as *Genyornis*. Their predators died of hunger.

In the six years between the two *Genyornis* papers, from 1999 to 2005, the Miller–Magee team saw megafauna politics heat up. The fire was fuelled by the Roberts paper of 2001, which is still inflaming passions today.

7 Frank the Diprotodon

In 1979, when jilleroo Louise Dunn (nee Friis) began excavating a skull from the bank of Cox's Creek, near Coonabarabran in northern New South Wales, she thought she was uncovering a piece of agricultural history—the remains of a bullock dating from the time when teams of the beasts hauled logs over the black soil plains.[1]

As the 21-year-old carefully removed soil from around the skull with tools she had seen archaeologists use on television, she realised this was quite a different animal. 'It had two very long bottom teeth that were about eight inches long, round ... with a chisel end,' she says. The molars were the size of matchboxes. 'The two front teeth were as long as the bottom ones and were one and a half inches wide.' She asked government geologists working in the area to examine the skeleton, which she had christened Frank. They thought the fossil could be a diprotodon, but urged Dunn to contact specialists to confirm the identification.

Dunn reached the Australian Museum's Alex Ritchie, an amiable Scotsman who looks like a palaeontologist from central casting and who is best known for his ground-breaking research on fossil fish and his campaign against creationism. 'He almost jumped down the phone when I told him what I thought we had,' she says. Before long, Ritchie and Australian Museum palaeontologist Robert Jones turned up to excavate the skull. The pair returned later with Sydney University archaeologist Richard Wright and a contingent of students and locals to liberate the rest of Frank from the bank. They removed 3 metres of overburden before getting to work on the precious skeleton with fine tools, wrapping each massive bone in toilet tissue before sheathing it in plaster to protect it on the trip to Sydney. (The toilet tissue makes it easier to remove the plaster.) They rushed to get the fossil out as flood waters rose, threatening it, and Dunn recalls the frustration of waiting for the plaster to dry.

Frank would become one of Australia's most famous diprotodons. In fact, he's the only one with a name. His skeleton was almost complete, and had an interesting feature—a puncture mark in one of the ribs. Wright was unable to say whether this was the elusive 'smoking gun'—a skeleton bearing a wound from a hunting weapon—so he did not publish details. Ritchie, for his part, is cautious in his interpretation of anything remotely related to a mammal, declaring he is really more of a fish person.

Frank, a member of the species *Diprotodon optatum*, would feature in countless photographs with Ritchie, all taken in the field from much the same angle. Frank the Famous Diprotodon, now on display at the Tourist Information Centre in Coonabarabran, would feature in the Roberts megafauna extinction dating project.

Diprotodon was described by Richard Owen in 1838 and named for its two front teeth, which give new meaning to the word 'prominent'. *Diprotodon optatum* stood about 1.5 metres tall on thick legs and was almost 3.5 metres long. Its massive head was built more to support its powerful jaws than its brain, which, at the size of a fist, was modest. Brains consume a lot of energy, and there was a bigger evolutionary advantage on a drying continent in harvesting bulk, low-grade food than in thinking. Whether the diprotodon was a browser or grazer is a subject of debate, but it was probably not fussy. It was once thought to be aquatic, a notion not as silly as it seems, considering the yapok, a semi-aquatic marsupial from South America that can seal its pouch. It is now known that the Australian beasts, not built for speed, lumbered over land on most parts of the continent. Curiously, no diprotodon remains have yet been found in Tasmania.

Their size might have proofed the adults, if not the young, against attacks from *Thylacoleo*, but the giant goanna *Megalania* might have posed a threat to all of them.

Frank was one of the last of his species. Soon after he died, something pushed all his clan into oblivion. When Louise Dunn discovered Frank's remains, radiocarbon dating, still hitting the barrier at about 40,000 years, remained the most reliable way of finding out when he lived. Frank was originally dated indirectly, through charcoal found near his skeleton, to about 38,000 years old. Other specimens seemed to be even younger—one of them, on the Liverpool Plains of central New South Wales, down to 7,000 years old. Scientists had trouble extracting enough collagen for radiocarbon analysis, so the specimens were dated indirectly, through charcoal or shell recovered from sediments in which they lay.

However, indirect dating is inherently unreliable because of the 'association' problem. Bones get moved around. Specimens can be redeposited by water or land slippage from older to younger sediments, so indirect dating of isolated bones cannot usually be trusted.

During the 1990s, Western Australian Museum palaeobiologist Alex Baynes reviewed 91 radiocarbon dates obtained for Australian megafauna from 42 sites. Using criteria set by US researchers David Meltzer and Jim Mead to assess American dates, he concluded that most Australian ages, including all those under 33,000 years old, were unreliable because of the material used for dating, or because the association between the date and the megafaunal remains was doubtful. He also considered the older radiocarbon dates suspect because they were nudging the effective limits of the method.

Richard Roberts's ambitious project to date the mass extinction using optically stimulated luminescence, to be published in 2001, was now under way. Roberts reunited with Rhys Jones and Mike Smith on the project and brought in other dating experts, along with palaeontologists. This was one of the first big projects to be led by a timelord, one of the custodians of the numbers, and it would send shock waves through Australian palaeostudies.

The work took in 39 specimens of 18 megafauna species from sites across the continent and New Guinea. The sites represented all prehistoric climatic zones—from the Centre through the temperate coastal fringe to the montane rainforests of New Guinea. Most remains from south-western Australia came from caves, while those from the east were found in swamps, along rivers, around lake basins and in coastal dunes. To home in on the extinction date, the 11-member team chose sites that were most likely to be the most recent, based on stratigraphy and geomorphology. That narrowed the field down to 28, the most that had been sampled in a single survey.

The scientists used a method that solves the association problem— they confined their analysis to complete or near complete articulated skeletons. To qualify, bones had to have been found in the sediments in more or less their correct anatomical positions, and with at least some still joined together. Whole skeletons would not have moved intact through the stratigraphic sequence, so the sand grains in which they were buried would give the right age. In contrast, jumbled bones could have intruded from older strata. The team spelled out its methodology in its Australian Research Council grant application and in the paper it later published in *Science*. When this reductionist approach elimi- nated some 'sacred cow' sites, opponents attacked the team for not publishing detailed stratigraphies.

The scientists indirectly dated 18 articulated specimens, returning to sites to get sand samples, including to the Cox's Creek bank where Frank the Diprotodon had been found. Rhys Jones had filled out by the time of the project, and he later recalled that he had to 'think thin' to get through the narrow passageways in Western Australia's Tight Entrance Cave, popping out like a 'human cork'.

The scientists also recovered sand grains from museum specimens. Team member Linda Ayliffe, of the ANU, confirmed many of the OSL results with uranium-series dating on limestone sheets, or flowstone, sandwiching strata containing megafauna remains at Mammoth Cave, Moondyne Cave and Kudjal Yolgah Cave in Western Australia. The team also dated a *Genyornis* footprint in a sandstone slab at Warrnambool, on the Victorian coast, and sediments bearing burnt *Genyornis* eggshell from Wood Point, South Australia.

Frank had made his exit just over 50,000 years ago, and mega- fauna from Mammoth Cave, first excavated by Ernest Lundelius in the 1950s, had died between 55,000 and 74,000 years ago.

To test the proposition that younger dates collected over the decades could be put down to the association problem, the scientists dated sand associated with several disarticulated specimens as well. Lending support to this hypothesis, many of these dates turned out to be young—the youngest, from Tambar Springs in New South Wales, was just 2,000 years. These and other young dates did not figure in the statistical calculation of the time of extinction, but formed the basis of one of the main points of the paper. The Cuddie Springs site in New South Wales, said by its excavators to have megafauna bones and artefacts in strata dated to between 32,000 and 38,000 years old, was also excluded from the statistical analysis—the remains were disarticulated, and the OSL dates on single grains of sand showed that the site had been disturbed.

None of the articulated remains was less than 46,000 years old. The youngest were a diprotodon and giant kangaroos 46,000 years old from Ned's Gully in Queensland, and another giant kangaroo, also 46,000 years old, from Kudjal Yolgah Cave. The scientists doubted they would be lucky enough to have captured the very youngest site in the survey, however. The oldest site was Victoria Fossil Cave in South Australia, which had dates of between 157,000 and 171,000 years for sediments sandwiching skeletons of *Zygomaturus trilobus*, *Thylacoleo* and *Procoptodon*. These pitfall trap victims had died long before human colonisation or the peak of the last Ice Age 30,000 years after colonisation. Including these and other very old dates in the extinction calculation would have skewed the results, giving an answer that was too old, so the team excluded sites older than 55,000 years from the final analysis. The logic of this was also spelled out in the *Science* paper.

The answer, based on a total of seven sites, was 46,400 years ago. It was close to the figure obtained in the *Genyornis* study by John Magee and Giff Miller, given the margin of error in the studies. The timing ruled out the last Ice Age as the cause of the extinctions. And the megafauna expired abruptly on geological timescales, eliminating long-term drying trends operating over hundreds of thousands of years as the cause. These trends are often used as a 'god of the gaps' in the extinction debate. The mass exit of big animals happened within 10,000 years of the arrival of people, suggesting a human role.

Detractors, led by the Cuddie Springs archaeologists and their supporters, ran hard on a few issues, including the number of sites included in the statistical analysis, as if the quantity of data was more important than its quality. In fact, the number of sites in the Roberts analysis exceeded that considered in the Cuddie Springs study by a factor of seven. The focus on articulated remains in the calculations 'rules out every archaeological site in this country', Cuddie Springs archaeologist Judith Field was quoted by *The Weekend Australian* newspaper as saying. 'They are excluding sites which don't fit their thesis.'[2] (In response to a question at the Legacy of an Ice Age conference, she later conceded that Cuddie was the only archaeological site affected.) Field, palaeontologist Stephen Wroe, then of the University of Sydney, and archaeologist Richard Fullagar ran with the issue in other media outlets, and Wroe tackled Flannery at the NMA public forum.

The statement in the *Science* paper that 'optical dating of individual grains from the Cuddie Springs deposit indicates that some sediment mixing has occurred' enraged some players. 'I'll tell you what is disturbed,' Fullagar was quoted by *The Weekend Australian* as saying. 'It is the sort of people who would publish such an interpretation with no collaboration with the most senior archaeologists who have been and are working on the site.'[3] Roberts responded with the observation that he had been the scientist who debunked Fullagar's claim to old dates for the Jinmium rock shelter in the Northern Territory—and added that the detractors were 'like an opposition party in parliament' who 'feel obliged to contradict us'.[4]

As with research into *Genyornis*, uncertainties in dating events in the remote past meant that Roberts and his colleagues could not tell whether the extinctions took a few hundred years or a few thousand. The data did not allow them to discriminate between a rapid hunting blitzkrieg or a slow burn. This conservatism won them no friends among their detractors, however. The scientists were attacked for not ruling out, or at least playing down, the blitzkrieg scenario, with one critic privately alleging political motivations.

The degree to which statistical data are scattered around the mean value—the centre of the famous 'bell curve', in this case, 46,400—is measured by the standard deviation. One standard deviation means the true value can be said with 68 per cent confidence to lie within a range tightly clustered around the mean. Statisticians are more confident—95 per cent sure—that the true value lies within a bigger range of two standard deviations from the mean. This is not a concept that lends itself to a 20-second media grab. The date of extinction to one standard deviation ranged from 43,600 to 48,900 years ago. To two standard deviations, it ranged from 39,800 to 51,200 years ago, putting it close to the date of colonisation and within striking distance of a blitzkrieg.

Roberts's team itself was divided. Some members, including Jones, Roberts and Smith, favoured human-caused environmental disruption (sitzkrieg), perhaps compounded by El Niño weather cycles, as the main mechanism of the extinctions. Flannery continued to back rapid overkill (blitzkrieg), arguing that the ages at the top of the range for human colonisation might be inflated.

The *Genyornis* and Roberts studies were backed up in 2007. That's when a team of palaeontologists and dating specialists led by Gavin Prideaux, of Flinders University and the Western Australian Museum, mined the fossil database in Cathedral Cave, part of South Australia's Naracoorte Caves system, to compile natural background extinction rate data. The Miller–Magee and Roberts studies had been labelled meaningless without long-range data on the natural extinction rate before people arrived. The Prideaux team filled that gap in the knowledge. The scientists found that the megafauna withstood ice ages for the half a million years before people arrived. The work, published in the prestigious US journal *Geology*, ruled climate out as the primary cause of the mass extinction.

The Prideaux team studied bones deposited in sediments in a chamber of Cathedral Cave over half a million years, a period spanning at least four ice ages. The mechanism of deposition had not changed over the period, so there was no sampling bias. Either the creatures—extinct and surviving species—fell into the caves and died there, or their bones were regurgitated by owls. The team's palaeontologists counted 62 non-flying species—and estimated their numbers. Among the megafauna, big kangaroos dominated, although other species, including *Zygomaturus* and the 'tree-feller' *Palorchestes* lived there too, among wombat and quoll species still living today. Little rodents reigned over animals under 5 kilograms. Roberts and colleagues, part of the Prideaux team, used OSL to date the sand entombing the remains.

The scientists compared their data with a 500,000-year climate record from Naracoorte stalagmites and stalactites in a previous study led by the ANU's Linda Ayliffe. As the speleothems form on cave ceilings and floors, they act as rain gauges, and Ayliffe used speleothem deposition to track rainfall and evaporation rates over several ice ages. Few sites in the world have such long, continuous, matching records. Although megafauna populations declined during dry periods, they rebounded, with most species withstanding repeated global climatic shocks. Larger animals declined during a dry period between 270,000 and 220,000 years ago, probably retreating from the region, but they returned, surviving well into the late Pleistocene.

The results tallied with another Naracoorte study, led by Donald Pate of Flinders University, on a megafauna deposit in Wet Cave. Pate's team, using radiocarbon dating on charcoal associated with the fossils, found that two giant kangaroo species, along with the predatory *Thylacoleo carnifex* and the hippo-like *Zygomaturus trilobus*, vanished from the region at least 45,000 years ago.

Despite a growing body of evidence against climatological causes, a few archaeological sites serve as rallying points for researchers contesting human involvement in the obliteration of the megafauna. Among them are Lancefield Swamp in Victoria, Seton rock shelter on Kangaroo Island, and Nombe rock shelter in the highlands of Papua New Guinea, but Cuddie Springs is the most controversial.

***

Extinct *Genyornis*, *Diprotodon* and *Sthenurus* are major characters in the Cuddie Springs drama, set on an archaeological site in the semi-arid zone of western New South Wales. But another megafaunal species, not native, not extinct and more of a villain than a victim in ecological debates, has a minor role, too. It is *Bos taurus*, the cow.

Megafauna bones were discovered at Cuddie Springs in the 1870s when local graziers dug a well to water their cattle. The site was excavated in the early 1930s by the Australian Museum and has been dug since 1991 by Judith Field and colleagues, including Richard Fullagar. Field, of the University of Sydney, did her PhD research on the site, one of the few in Australia with megafauna bones and stone tools in the same levels. The Cuddie Springs team says some of the bones have 'modifications that are consistent with butchering and burning'. It says the site 'rebuts' the Roberts results, 'refutes' blitzkrieg, and supports the climatic model of mass extinctions.

The critical stratigraphic unit with artefacts and megafauna bones— unit 6—lies between 1 and 1.7 metres below the surface of a claypan in the centre of what the research team describes as a 2-kilometre- diameter ephemeral lake. The team says unit 6 is sealed above by a 100-metre-diameter, 5-centimetre-deep former land surface, called a 'deflation pavement', which also has megafauna bones, along with stone artefacts at an extraordinary density. Cow bones are mixed with extinct megafauna bones in the sediments above the deflation pavement, but not those in or below it, and the unit is sealed below by another ancient land surface.

Attempts at dating the megafauna bones directly through the radiocarbon and amino acid racemisation (AAR) techniques failed, but the team got radiocarbon dates on charcoal in the sediments bearing the bones. The numbers for the sealed layer, unit 6, ranged from about 32,000 to 38,000 years, younger than the mass extinction dates obtained in the Roberts and Miller–Magee studies and younger than the most conservative dates for human colonisation of Australia. The Cuddie team says the remains are from animals that became trapped in the mud and died naturally, or were scavenged and butchered by humans. The latter scenario, which would account for the lack of articulated remains at the site, has received more airplay.

The team says grindstones, resembling grinding tools found elsewhere in late Holocene sites and indirectly dated to more than 30,000 years old, are the 'oldest direct evidence for processing plant materials in Australia'. It also claims that blood residues and hair on stone tools are evidence for butchering. Previously, the team has said that the late Tom Loy, who features in the book *Jurassic Park*, recovered diprotodon DNA from tools, but that work was never formally published.

Some scientists, including most timelords, contend that the site is disturbed. The 2001 Roberts results that had such an incendiary effect on the Cuddie group were based on the dating of single grains of sand from the top pavement and unit 6. Individual grains from the same levels should have come in at the same age but did not. The Roberts team interpreted this, and the younger ages for the disarticulated remains, as evidence that the bones had been eroded from older units and redeposited in younger sediments.

In 2006, one of us, Richard Gillespie, and the University of Adelaide's Barry Brook fanned the flames with publication of a comprehensive analysis of the Cuddie dates, suggesting that the deposit was disturbed. Gillespie examined bones and sediments from the site while Brook did a statistical analysis on 16 published radiocarbon dates, one from the top pavement and the rest from the crucial unit 6, to test the proposition that the unit was an intact stratigraphic sequence, as the Cuddie Springs team maintains. If it were, the ages should get older with depth, but Gillespie and Brook found that when uncertainty in the dates was taken into account the charcoal was all much the same age—about 36,000 years. They noted that the site contained an unusually high concentration of charcoal—an estimated 3.5 tonnes in the deflation pavement, and probably more in the top sediments of unit 6. That, they said, was too much for a campsite used only intermittently when the lake was dry, and the charcoal was more likely to have been dumped into the deposit by floodwaters after bushfires. Excavations had so far turned up no hearths, ovens or fireplaces, they added.

Gillespie and Brook agreed with the assessment of Field's team that the artefacts resembled Holocene assemblages found on many other sites. In their view, the tools were more likely to date from that period. That, they said, explained the blood and hair found on some of them. Gillespie's attempts to extract collagen from 10 bone samples, chosen by Field, found the protein in only one—a modern cow bone. Hair, said Gillespie and Brook, was even less likely than bone collagen to be preserved in the seasonally wet open site.

The two echoed comments made by archaeologist Bruno David amid much controversy in 2001 that the upper pavement at Cuddie Springs might have been created in the late nineteenth or early twentieth century to 'create a firm footing for cattle'. They speculated that the pioneering graziers probably saw the tools as 'gravel of convenient size' and brought it in on carts drawn by bullocks or horses. And in any case, the site was probably in a palaeochannel—the bed of a Pleistocene river that transported the charcoal and bones to the region. They cited as evidence supporting their interpretation a 54-metre-deep core, drilled by the New South Wales Geological Survey just 60 metres from Field's dig, which turned up no bones or artefacts. And the concentration of coarse sands and gravels down the profile of the site suggested water rather than wind deposition, they added.

How did the artefacts get jumbled up with the bones? Trampling by cattle could be one explanation. Disturbance from the well digging and museum excavations could be others. The Cuddie team published research in *Proceedings of the National Academy of Sciences* in 2005 comparing the rare earth element (REE) signatures in bones from the same levels. The researchers said that REEs were adsorbed 'rapidly from pore waters' onto the bone surface crystals and locked in the crystal lattice. Post-depositional mixing would show up as bones with different REE prints, but the readings of the elements—lanthanum, samarium, ytterbium and cerium—in bones from unit 6 suggested the remains were in situ. Gillespie and Brook argue that the results do not rule out an origin of the bones beyond the claypan, since the study compares the bones with each other, not with their environment.

The debate replays old arguments about other sites held to demonstrate a long overlap of people with late-surviving megafauna. Gillespie and Brook went on the offensive with a press release setting out their reinterpretation. Field parried with the observation that Gillespie and Brook were not archaeologists, and had never been to Cuddie Springs. She was quoted by *The Canberra Times*: 'To suggest, firstly, that you can use the statistical analysis of dates to demonstrate the site is disturbed is absurd ... To suggest there's sediment disturbance on the basis of a dating study is completely off the planet.'

Fullagar conceded to *The Australian* newspaper there was 'some interesting discussion on radiocarbon dating statistics', but said 'the explanation they put forward is fundamentally flawed. It's wrong ... you've got sediments where Aboriginal people have been camping for long periods of time, the artefacts are going to be scuffed around and moved around a bit.'

Megafauna bones were discovered in 1843 at Lancefield Swamp, on the outskirts of a small country town 70 kilometres north of Melbourne, by a well-digger. Further investigation was prevented for more than a century by the high water table. Only in the 1970s, with the use of pumps, did archaeologist David Horton and his team begin large-scale excavations. Gillespie, then at the University of Sydney, did the radiocarbon dating work. He obtained dates of 3,000 to 24,000 years for megafauna bones but always doubted the results. 'I did then what people did in the seventies, which was to date everything that didn't dissolve in acid,' he says, adding that Polach had used similar procedures when dating Mungo Lady. Scientists were only just beginning to appreciate the subtleties of the method at the time. Gillespie now says he had probably dated humic acids taken up by the bones from the burial soil. Charcoal from a flood channel beneath the bone bed came in at about 30,000 years old, but Gillespie has long since retracted those results too, because the bones were not directly associated with either the charcoal or the few artefacts found.

In 1998, a team led by Sanja van Huet of Monash University published data supporting the view that the Lancefield megafauna remains were older than previously claimed. The scientists got an ESR age of about 50,000 years for a diprotodon tooth in one of the three megafaunal deposits at the site. Radiocarbon dates on apatite from teeth came in at just over 30,000 years, but the authors stressed that 'apatite dates always represent a minimum age'. AAR dating on a tooth yielded an age of between 30,000 and 55,000 years.

The site continued to be held out by some as proof of a long human–megafauna coexistence. Horton continued to quote the original numbers into the new millennium. 'In a nutshell, the megafauna are dated at less than 30,000 at Lancefield, together with indications of environmental change and circumstances that suggest a mechanism,' he wrote in his book, *The Pure State of Nature: Sacred cows, destructive myths and the environment*, published in 2000. 'Every effort to shake the date from this site has failed, and I have heard that it has recently been confirmed.' Some researchers were still barracking for Lancefield at a 2001 National Museum of Australia symposium for researchers, and later, but preliminary OSL dating results presented at the 2005 Quaternary Extinctions Symposium put all the megafauna there at more than 40,000 years old.

Another site said to challenge human agency in the extinctions is Seton rock shelter on Kangaroo Island off South Australia. It was excavated by Jeanette Hope, who describes it as a Tasmanian devil den occasionally used by people. But many researchers argue that the megafaunal remains there—three *Sthenurus* tooth fragments, one in a layer dated to 19,000 years ago and one each in levels immediately above and below it—do not amount to enough material for them to sink their teeth into. Meanwhile, Nombe rock shelter in PNG, also said to be around 19,000 years old, has no articulated remains.

Gillespie, Barry Brook and Alex Baynes have analysed published dates for megafauna and archaeological sites in a study aimed at solving the GIGO (garbage-in, garbage-out) problem. They wanted to find the period of overlap of people and megafauna in a bid to discriminate between human and climatic causes, scrutinising dates gathered over decades and etched into the record—databases still in use. They eliminated many of the dates on the grounds that they were based on outdated chemistry and carbon-14 measurement technology. Others, as Baynes had found in earlier research, did not meet criteria set by David Meltzer, one of the leading proponents of the climate-caused extinction model, and palaeoecologist Jim Mead. Those criteria scored radiocarbon dates from American megafauna sites according to the type of sample measured (with charcoal getting top billing) and the degree of association between sample and extinct animal (with direct dates on bones or dung favoured). Basing their study on more stringent criteria grounded in recent research, Gillespie, Brook and Baynes axed more than 100 radiocarbon dates from the Australian megafauna databases, and the team used results only from OSL and uranium-series dating for megafauna. Combining these with the archaeological record based on the most reliable charcoal radiocarbon dates, they calculated a human–megafauna overlap of about 3,900 years—a small enough number to suggest human agency and, given the dating uncertainties, within the bounds of a blitzkrieg. Climate change was ruled out.

In 2007, a 14-strong international palaeo-SWAT team, including Flannery and Martin along with top geologists, model-builders, ecologists and dating specialists, asked what had become the blindingly obvious question: 'Would the Australian megafauna have become extinct if humans had never colonised the continent?'

The question was aimed squarely at a review by Wroe and Field of the Australian data on climate and extinction, and the two were ready for a stoush, somehow managing to get their response up on the *Quaternary Science Reviews* website before the SWAT team's paper had appeared in the journal. They concluded: '... we remain confident that despite enthusiastic attempts to shut down the debate, the respective roles of climate and humans will be the subject of robust argument for a long time to come.'

The SWAT team pointed to the science already in the peer-reviewed public domain. 'Globally,' the authors said, 'heavy extinctions in the Late Pleistocene (Australia), terminal Pleistocene (America), Early to Middle Holocene (West Indies and Mediterranean islands) and Late Holocene (Madagascar, New Zealand and Pacific Islands) always coincide with human colonisation. Recent work has demonstrated clearly, yet somewhat paradoxically, that neither a specialised tool kit for hunting big game, nor mass-kill sites are required, or even expected, under a "blitzkrieg" model of overkill.' The team also suggested which new science might extend existing knowledge.

Many scientists had by now moved on. The research on eggshell, bones and charcoal had put people squarely in the frame for the Australian megafauna extinction, with local El Niño droughts perhaps compounding human impacts. The SWAT team suggested, echoing comments by the University of Arizona geologist Vance Haynes in the American debate: 'Genuinely "falsifiable" hypotheses should be used to advance understanding and reduce uncertainty, while stimulating debate and fostering the development of new ideas and innovative tools.' The team concluded: 'In sum, the question is no longer if, but rather how, humans induced this prehistoric extinction event.'

A big ESR dating project by Rainer Grün, Flinders University palae-ontologist Rod Wells and colleagues set out to falsify the blitzkrieg hypothesis, but failed. The team went out looking for sites younger than the 51,000- to 40,000-year extinction window calculated by the Roberts team. Flannery had offered boutique red wine as a prize if they could pull it off. The scientists directly dated marsupial teeth, mobilis-ing the ANU's new technique, prosaically named 'multi-collector laser ablation inductively coupled plasma mass spectrometry'. (The team is still working on an acronym for the acronym, MC-LA-ICPMS.) The method helps in the calculation of a critical parameter—the sample's internal radiation dose rate, which depends on the concentration of natural radioactive isotopes, especially uranium, that power the ESR signal.

The sample is moved across a pulsed laser, which blasts holes with dimensions on a scale of a few tenths of a millimetre. The laser beam, operating in the ultraviolet band, would damage your eyes if you looked at it. It detonates mini-explosions, atomising material and leaving tiny, almost invisible craters as it steps across the sample in small increments. The material blown off the sample is channelled to a plasma flame which heats it to 7,000 degrees Celsius, ionising it before it is passed to a mass spectrometer. The readings feed into the internal dosimetry calculations for the ESR dating and provide uranium-series dates as well.

The teeth came from six megafauna sites in a north–south transect from Hookina Creek in South Australia's centre, through the Flinders Ranges and on to Kangaroo Island. Sites were chosen to span a range of natural rainfall and other climatic indicators. The team reported in the *Australian Journal of Earth Sciences* that the samples were all older than 40,000 years.

Further, support for the blitzkrieg scenario would come from an unexpected source—cyber-megafauna.

8 Silicon beasts

A 'two-tonne wombat' wanders past. But rather than the adult diprotodon, it's a nearby juvenile that captures the attention of a hungry clan fed up with chasing bilbies.

The young diprotodon will be easy to take, even without big-game hunting weapons. It is 500 years since a founder population of 50 to 100 people made landfall in Australia, but people are still living in low densities—one person per 10 square kilometres. The megafauna they hunt opportunistically has been in decline, but knowledge of how many big animals were around in the old days has gone from the corporate memory. The kill rate is only one or two juvenile diprotodons per large family group per year, but soon the big marsupials will vanish. This virtual palaeoworld was generated on a computer by ecologists Barry Brook and David Bowman, of the University of Tasmania, and Chris Johnson, of James Cook University, in two studies aimed at testing the blitzkrieg hypothesis.

The scientists compiled a database of body masses for 198 extinct and 433 surviving species weighing over 5 kilograms from Australia, Africa, Eurasia, North and South America and Madagascar, basing estimates of the values for extinct fauna on fossil bones. They developed a population model—a set of mathematical equations representing population dynamics—to investigate possible scenarios for the mechanism of extinction. The 150 scenarios were generated by varying values assigned to five parameters—human population density, prey population density, replacement rate, kill rate and megafauna naivety.

Other parameters could be factored in to the five basic ones: the targeting of juveniles, for example, could be reflected in the value assigned to the prey replacement rate, and human- or climate-induced habitat destruction could be expressed in the prey population density.

The scientists drew on archaeological and ethnographic data on hunter-gatherers to reconstruct Pleistocene demography, basing their input data for extinct megafauna on knowledge of comparable living species. They treated the diprotodon as a scaled-up version of its closest living relative, the common wombat, using a lifespan of 30 to 40 years, with females first mating at six to eight years old and reproducing only once every three years. They used the scenarios most closely fitting the known relationship between body mass and extinction risk to calculate the rate of extinction.

The result? Several scenarios, including those involving a reduction in the prey replacement rate due to the targeting of juveniles, tallied with the body mass/extinction relationship. Other scenarios did, too, so the scientists were unable to pinpoint the exact mechanism of extinctions. However, all the simulations closely fitting the relationship had one thing in common: the prediction that the giant animals became extinct within 800 years of human colonisation of continents, and sooner on islands—blitzkrieg. Scenarios without hunting failed to generate megafauna extinctions, suggesting that neither climatic factors nor anthropogenic burning could alone have triggered the catastrophe.

The speed of extinctions surprised the scientists; Bowman once doubted human agency in the mass extinction. And the modelling showed how even a small harvest could have drastic consequences. But, in keeping with earlier modelling work, the rate of extinction was slow enough to be imperceptible to people on the ground. Brook likens the results to the sharp decline in wandering albatross numbers in the latter half of the twentieth century, a crisis that took everyone by surprise. And modelling done by CSIRO's Geoffrey Tuck and colleagues showed how vulnerable that species could be. It predicted that the loss of just over 40 adult female albatross a year as by-catch in long-line tuna fishing in the Southern Ocean could be enough to threaten the Crozet Islands colony in the Indian Ocean.

Ecologists developed population models in the 1970s as wildlife management tools. The models were soon applied to prehistory, with Paul Martin and colleague James Mosimann in 1975 publishing results of a simulation sketching the rapid colonisation and megafauna extinction of the Americas. The most comprehensive modelling work so far, done by John Alroy of the University of California, caused a stir when it was published in *Science* in 2001 alongside the Australian study by Richard Roberts and colleagues. His model correctly predicted the fate of 34 out of 41 species of North American herbivorous megafauna, some that expired in the mass extinction and some that made it through, according to a literature survey by Anthony Barnosky, of the University of California, and colleagues. The median time to extinction was 895 years. The simulations, Alroy wrote, demonstrated that overkill was not only plausible but unavoidable. 'The overkill model thus serves as a parable of resource exploitation, providing a clear mechanism for a geologically instantaneous ecological catastrophe that was too gradual to be perceived by the people who unleashed it.'

Other simulations designed to test the blitzkrieg hypothesis have had mixed results, depending on input parameters and on whether the prey and human population models were set up to interact, reflecting feedbacks as numbers grow or decline, says Barnosky's team. But how close to reality are the ancient worlds created in silicon chips? The models are being refined, and many are validated in tests on historical data. The biggest drawback is in the veracity of input data based on educated guesses, but some of those numbers will harden up. Hints on megafauna reproduction rates could come from the growth rings in teeth, which record when females start reproducing and when they are pregnant. Ancient DNA could yield hard data on the population dynamics of the extinct fauna, too—but probably not on mainland Australia, where DNA is unlikely to have been preserved.

Other research supports the ecosystem destruction mechanism of extinction. The first hard evidence came from a volcanic crater.

***

Viewed through a microscope, the structure shielding the sperm cells of plants is an object of great beauty. The pollen grains of eucalypts are triangular, while those of Australia's acacias, the wattles, resemble soccer balls. Hoop pine pollen grains are spherical, while the Poaceae, the native tussock grasses, have rounded grains with a distinctive aperture. Intricate patterns are etched into the surface of the grains of many species. The grains, which measure between 10 and 100 millionths of a metre, protect their precious contents with a sheath of sporopollenin, one of the most durable organic materials on the planet. They waft into oceans, lakes and swamps, settling in sediments where their outer walls can last for hundreds of millions of years. To reconstruct past landscapes, palynologists read this botanical archive through microscope lenses set on 200 to 1,000 times magnification.

Ancient pollen from Lynch's Crater, a volcanic remnant on the Atherton Tableland in north Queensland, holds some of the answers to questions about long-term landscape change in Australia. The volcano was last active about 250,000 years ago, after which a lake formed in the crater. Since about 50,000 years ago, it has been more like a peat swamp. The sediments hold northern Australia's longest and most complete pollen record and charcoal particles that reveal the region's fire history—a continuous archive spanning two ice ages. And Lynch's Crater was one of the first terrestrial sites in Australia to be matched with a sediment core from the seabed, providing a cross-reference for data from the land. Together, the records open the possibility of teasing out the effects of forces acting on the landscape with differing strength and over differing timescales, driving the transition from rainforest to fire-loving sclerophyll bush dominated by eucalypts.

Peter Kershaw, a palaeoecologist at Monash University, has been studying the Lynch's Crater record for three decades, and published the definitive paper on the site in *Nature* in 1986. The pollen paints a picture of an ancient lake surrounded by towering araucarian conifers like the Norfolk Island and hoop pines, once widespread across the globe, but which had retreated to southern refuges by 65 million years ago. Another group of southern pines, the podocarpus conifers, which had evolved on the ancient southern supercontinent Gondwana, also graced the lake. This suite of plants dominates what is now called 'dry' rainforest, and the southern pines are extremely fire sensitive.

A spike in charcoal particles in the Lynch's Crater sediment core, radiocarbon dated to about 42,000 years, marked an upsurge in bushfires, and the pollen recorded a change to sclerophyll bush, mostly eucalypts, acacias and casuarinas. The sclerophyll claimed the region until humid conditions after the end of the last Ice Age 12,000 years ago. The moister climate of the present interglacial period set the scene for an expansion of 'wet' rainforests into the region, as it had in previous interglacials. These forests of figs, red cedar and laurels now share the area with the sclerophyll bush, but the southern pine forests have never returned.

When the radiocarbon date came in at 42,000 years, Kershaw thought he might be reading the story of the environmental impact of fires lit by the first Australians. Most archaeologists thought that humans had first reached the continent around then. In 1986, however, 42,000 years was right at the radiocarbon barrier, so in 2001 a team led by Chris Turney, then at Wollongong University, and including Kershaw published research redating the site. The team had drilled another core and tried the ABOX-SC sample pre-treatment method on the sediments. Finding that little survived the chemistry, and the ages too scattered, they reverted to conventional acid–base–acid treatment, and published their results in the *Journal of Quaternary Science*. The level at which charcoal particles increased, starting from 11 metres down, turned out to be about 50,000 years old, although the ages reached another radiocarbon limit there. The results tallied with those from the ODP-820 marine sediment core, a 68-metre-long sequence spanning 250,000 years. The core, from the continental slope near the Atherton Tableland, was recovered by the bluewater drillship *Joides Resolution* as part of the international Ocean Drilling Program. It had high charcoal levels at 130,000 years ago, just after the second-last ice age, and again from 50,000 to 30,000 years ago.

The 50,000 figures from Lynch's Crater and the sea were close to the most recently calculated dates for human colonisation. And the fire regime then differed from that in the ice ages, implying a human role. There had been no comparable increase in burning and vegetation change around the harsher times of the penultimate ice age, from 130,000 to 180,000 years ago. The vegetation had changed gradually after the start of the big burn, over 20,000 years, whereas transitions driven solely by the ice ages had occurred abruptly. The big question was the role in the vegetation shift of droughts caused by the El Niño/ Southern Oscillation (ENSO). Kershaw's reading was that human impact was probably minimal until El Niño kicked in.

In 2004, Turney, Kershaw and colleagues published palynological research in *Nature* using the battle for supremacy between sedges, favoured by wet conditions, and grasses, to gauge the impact of El Niño at Lynch's Crater. The scientists also used the decay of peat, which is accelerated by droughts, as a proxy for rainfall change driven by ENSO fluctuations. They detected periods of severe El Niño droughts at 11,900 and 1,500-year intervals over the past 50,000 years, augmenting the youngest part of John Chappell's coral record from New Guinea.

Chappell's team had measured oxygen isotopes in 30- to 100-year slices of coral, using it as a southern oscillation index of the past. The scientists got seasonal-scale information on sea surface temperature and salinity through two glacial cycles. They found that ENSO had been operating for at least 130,000 years but had been strongest in the current Holocene interglacial. Compared with today, ENSO variability was lower, and the climate more stable, when humans first arrived in Australia near the start of the OIS 3 interstadial. The coral study gives high-resolution snapshots of natural climate variation, while the lake sediments overlay 50,000 years of ENSO variability on widespread landscape burning attributed to the first Australians.

ANU palaeoecologist, the late Gurdip Singh, claimed in 1985 that humans could have been responsible for 130,000-year-old charcoal peaks in sediments from Lake George, near Canberra. Sclerophyll woodland pollen in the lake cores broadly matches the Lynch's Crater record, although the dating beyond radiocarbon range is still uncertain.

Meanwhile, researchers studying lake cores from New Caledonia disagree with the landscape firing interpretation. The ANU's Geoff Hope and Janelle Stevenson have argued that because humans did not reach New Caledonia until about 3,500 years ago, it should be possible to disentangle natural from human-induced changes. They found a decline in araucarian rainforest at Lake Xere Wapo, similar to that in Lynch's Crater, about 50,000 years ago. Since human disturbance could be ruled out on New Caledonia then, they concluded that climate change was the more likely cause of the vegetation change in both records. The chronology of the New Caledonian core is controversial, however.

Some scientists say it is impossible to differentiate between human-lit and natural fires through the analysis of charcoal particles in core sediments, so it is speculative to invoke a human role. They say Rhys Jones's 1969 'firestick farming' hypothesis was based not on archaeological evidence but on nineteenth- and twentieth-century ethnographic observations projected 50,000 years back to first landfall. Others disagree, arguing that the conclusions on Lynch's Crater centre on the palynological evidence for differing fire regimes. And how would firestick farming show up in the material culture of the archaeological record?

Other signs of ecosystem disruption have come from the interior.

***

A team of scientists led by Flinders University's Gavin Prideaux studied marsupial fossil teeth from caves in the Nullarbor Plain. Discovered by cavers in 2002 in one of Australia's biggest fossil finds ever, the remains comprised 69 vertebrate species, both extinct and extant, including 23 kangaroo species, eight of which were new to science. Many species of extinct megafauna were among animals that had plunged to their death in the caves through pitfall traps. Their skeletons were among the best-preserved ever found, and the discovery drew widespread publicity. One specimen is the first whole skeleton recovered of the 'marsupial lion' *Thylacoleo carnifex*—the biggest carnivorous marsupial. It could force a rethink of the lifestyle of the amazing predator. The paper on the find, published in *Nature*, also describes a snail, but *Pupoides adelaidae* has not attracted much media attention.

The team deployed three techniques to date the fossils. John Hellstrom, of the University of Melbourne, did uranium-series dating on mineral crusts formed on the surface of the bones. The Australian National University's Brad Pillans used palaeomagnetic dating, a method based on the signature in some rocks and sediments of reversals and excursions in the direction of the Earth's magnetic field, dated by other techniques, while Richard Roberts and colleagues did OSL dating. The bones had accumulated in the deposit sometime between about 800,000 and 200,000 years ago, long before humans arrived on the continent.

To see what the climate was like when the Nullarbor animals lived, the ANU's Linda Ayliffe analysed tooth enamel from 13 kangaroo and one wombat fossil for oxygen-16 and oxygen-18. These isotopes betray rainfall and evaporation rates. Water molecules with the lighter isotope of oxygen are the first to evaporate, so dry areas become enriched in the heavier form of the element. Ayliffe compared the results with those of modern kangaroo and wombat roadkill, as well as museum specimens, from regions of known rainfall levels around southern Australia. Rain in the cooler, wetter Hampton Tableland to the south is delivered by the winter westerlies. The drier, hotter region to the north is dominated by the summer monsoon. The oxygen isotope numbers for the ancient teeth fell between the modern results for these regions, suggesting, in general, a dry climate similar to today's, with the Nullarbor getting about 200 millimetres of rain a year and no strong seasonal rhythm.

However, a rollcall of the fossils painted a picture of a past landscape vastly different from the saltbush wasteland of the Nullarbor today. Grazers, like the giant wombat *Phascolonus gigas*, dominated the landscape, suggesting vast areas of open country. But the region also supported nine species of browsers, or leaf-eaters, including two new species of tree kangaroo and the extinct *Procoptodon*. 'To support this biodiversity, the region must have been a mosaic of grasslands and woodlands,' says Prideaux, adding that fire-sensitive trees, like the quandong, with palatable leaves and fleshy fruit, probably broke the monotony of the mallee scrub and wattles.[1]

If the rainfall in these ancient times was similar to today's, what transformed the landscape? The scientists speculate that fire was the most likely culprit. They also suggest that the Nullarbor discoveries bolster the case for habitat destruction as a contributing factor in the megafauna extinctions 46,000 years ago, while stressing that direct human hunting may also have contributed. 'We've got the vegetation changing, meaning it's unlikely to be a purely hunting scenario if overhunting was involved at all in the extinctions,' says Richard Roberts.[2] 'It's likely there was some juvenile predation, but given that there's now evidence of a human role in the firing of the landscape, the notion of a marauding band of Aboriginal hunters is unnecessary.'

Further work on eggshell has produced results analogous to the Nullarbor study, and the Miller–Magee team drew similar conclusions to the Prideaux team's. The Miller–Magee team compared oxygen isotope values for carbonate in *Genyornis* and emu eggshell before the *Genyornis* extinction, 50,000 years ago, with those from emu eggshell after it, 40,000 years ago. The source of the oxygen was the birds' drinking water. Miller told the International Union for Quaternary Research conference in Cairns in 2007 that the results suggested only a gradual drying down from 60,000 years ago, so climate could not account for the abrupt vegetation change between 50,000 and 45,000 years ago recorded in the eggshell carbon isotopes. 'Human burning transformed the vegetation across semi-arid Australia,' said Miller.

The Lynch's Crater pollen, the teeth from the Nullarbor fossils and the *Genyornis* and emu eggshell suggest a continent engulfed by fire.

Australia has been seen as a natural laboratory that controls for major global climate change, enabling the decoupling of human from climatic impacts. It could shed light on the American extinctions. They happened while the ice sheets that covered much of the northern continent were retreating as the Ice Age loosened its grip ahead of the Younger Dryas, a sharp cold snap. But perhaps research in America can give hints on the Australian scene, too.

***

The Australian battles over the date of colonisation and of megafauna extinctions are mirrored in the New World, even though it was not considered *terra nullius* by the conquistadores. 'They ought to make good and skilled servants, for they repeat very quickly whatever we say to them,' Christopher Columbus wrote in his journal on October 12, 1492. 'I think they can very easily be made Christians, for they seem to have no religion.' More than 500 years later, scientists and archaeologists argue about when the ancestors of these 'savages' colonised America, their role in the local megafauna demise and the effect of nature—climate change and cosmic impacts. Genes, dung, glass beads and buckyballs figure prominently in the debate.

9 New World order

Silvia Gonzalez is no shrinking violet when it comes to the date of colonisation of the Americas. According to a BBC report, she described her claims of dates of 40,000 years for human footprints from Mexico as an 'archeological bomb', declaring that her research team was 'up for a fight'.

Gonzalez, of Liverpool John Moores University in the UK, and colleagues got the 40,000-year estimate for the site in the Valsequillo Basin near Mexico City through radiocarbon dating on freshwater shells, ESR dating on mammoth teeth and the OSL method on quartz sand associated with the impressions, said to be made in a layer of hardened volcanic ash.

Gonzalez, who has argued previously that the first Americans could have been seafarers from Australia, challenges the 'Clovis First' orthodoxy. According to this, the 'Clovis hunters' entered North America via Beringia—an ice-free thoroughfare that stretched from north-eastern Siberia, across the Bering land bridge to western Canada—no earlier than 13,000 years ago. The claims for the footprint site met with scepticism about the dates and the interpretation of the impressions. They got widespread publicity at a London press conference in 2005 during the Royal Society's Summer Science Exhibition. But before the Gonzalez team had published its results in *Quaternary Science Reviews* it was gazumped by Paul Renne and colleagues, of the Berkeley Geochronology Center in California. That team used argon–argon dating and palaeomagnetic measurements to claim that the ash layer bearing the footprints was about 1.3 million years old. The age pre-dated by more than a million years the earliest uncontested dates for the arrival of humans in the Americas and for the first known *Homo sapiens* in Africa. Renne's team concluded in *Nature* that the 'footprints' had been wrongly identified.

Gonzalez had previously claimed star status for Penōn Woman III, a skeleton also found in Mexico. The claim was based not on the radiocarbon age of the skeleton, which at 12,700 years old was prime-time Clovis, but on skull measurements. Gonzalez said Penōn Woman's skull, like several others from Mexico, was longer and narrower than most Amerindian skulls. Such features, she suggested, were more European than north-east Asian.

The American colonisation debate alternates between hard science and radical pseudoscience. Academics argue about the Clovis First hypothesis and about the route the colonists took south—a trek along an ice-free corridor between the ice sheets that covered much of North America during the last Ice Age, or navigation of the Pacific coast? The academics, meanwhile, have been accused of a cover-up. Michael Cremo and Richard Thompson claim in their book *Forbidden Archaeology* that academic archaeologists and geologists have suppressed details of hundreds of modern human skeletons from multi-million-year-old sediments in America and Europe because the remains do not fit standard theories. And like their Australian counterparts, some indigenous Americans claim their people have 'always been here', so any old age suits them, and the older the better. Meanwhile, a chorus of creationists has condemned all dates older than 4004 BC.

The debate has raged since the discovery in 1927 of fluted projectile points associated with the skeletons of extinct bison near Folsom, New Mexico. The find underlined the antiquity of human occupation of North America. Extinct animals were thought to be Pleistocene age, so people must have entered America then, not more recently, as the archaeology known at the time suggested. The find was followed in 1934 by the discovery of bigger fluted projectile points near Clovis, also in New Mexico. Clovis points came from deeper, and therefore older, deposits than Folsom points, and several other styles of projectile point have since been described from younger deposits above the Folsom layers.

Harvard geologist Kirk Bryan noted in 1941 that the Folsom and Clovis archaeological discoveries had sharpened the focus of geologists, particularly on landscape changes caused by the advance and retreat of glaciers. Because the fluted points were associated unequivocally with bones of extinct animals—bison in Folsom levels, and mammoth in Clovis levels—scientists were chasing a dream of correlating the archaeological stratigraphy to 'a general geologic chronology in step with the rhythm of climatic fluctuations'. While admitting that the geologic method for dating was 'full of pitfalls and blind alleys', Bryan observed wryly that it was the only game in town.

Estimating the ages of Clovis and Folsom cultures involved a chain of geological inference. First, researchers had to establish the stratigraphy of the archaeological sites and correlate them with fossil sites containing the same extinct megafauna species as those associated with the different fluted points. They then had to correlate these scattered sites with the general North American glacial stratigraphy, then still under development. Sites such as Lindenmeier in Colorado and Hell Gap in Wyoming were close enough to trace the layers bearing artefacts up streams and gullies to the nascent glacial stratigraphy of the Rocky Mountains. This was tied to glacial sites 1,500 kilometres away in the Great Lakes region, and from there a 5,000-kilometre leap of faith to the far superior European glacial chronology established by counting varves (layers of lake sediments).

Bryan's best guess was that Folsom points were between 25,000 and 10,000 years old, and he favoured the older end of the range. Ernst Antevs, of the University of Arizona, thought the younger end was more likely, perhaps about 13,000 years ago for the Lehner site in Arizona, where Clovis points and mammoth remains had been found. These dates were based on distant, supposedly annual, records—absolute numbers, before the advent of radiocarbon dating.

In the 1970s, Jeff Bada, of the University of California at San Diego, obtained amino acid racemisation (AAR) dates of up to 70,000 years for human skeletons excavated from coastal sites in California. Was there really 'crypto-archaeology' that decades of careful excavations had missed, only now revealed through this new dating method? Bada later retracted the results when independent dating at the Oxford accelerator mass spectrometry facility showed the skeletons were less than 12,000 years old, with some much younger.

The Oxford team, which included one of us, Gillespie, had tested stringent new decontamination methods on control samples of known age, including a pig bone recovered from Henry VIII's flagship *Mary Rose*, sunk in action against the French in AD 1545, and a human bone from the ruins of Pompeii, buried by volcanic ash from Mount Vesuvius in AD 79. The Oxford lab also processed samples of animal bones from geological deposits, including some from a 35,000-year-old extinct woolly rhinoceros.

Archaeologists who had questioned the racemisation ages and correctly interpreted as Holocene the style of artefacts found with the human bones had been vindicated. And if the human bones really had been 30,000–70,000 years old, it would have been difficult to explain shells and other debris of a maritime culture, because sea level was much lower then and the coastline would have been several kilometres to the west. Many were disappointed because the new radiocarbon dates supported neither the short (biblical) nor the long (indigenous) chronology.

Scientists have since developed AAR dating, and the method is now well established.

Calico Foothills is among other archaeological sites north of the Rio Grande for which old dates have been claimed. Louis Leakey championed the site. He thought artefacts found there resembled Oldowan tools, the first technology developed by humans—more than 2 million years ago—and named for the Olduvai Gorge in Tanzania, where they were found by Mary Leakey in the 1960s. Geologists had assured Leakey that the Californian site was 50,000 to 80,000 years old. Most archaeologists now agree that the putative tools are geofacts, not artefacts.

And an acrimonious debate over radiocarbon results for the Meadowcroft rock shelter in Pennsylvania remains unresolved. James Adovasio, who excavated the rock shelter for 30 years, claims occupation 19,000 years ago, when the margin of the Laurentide ice sheet was just 80 kilometres north of the site. The published dates are controversial, and Adovasio has refused to have seeds and other botanical remains measured.

South America has several claims for pre-Clovis occupation, the most persistent being for the Monte Verde site in southern Chile. Tom Dillehay, of the University of Kentucky, maintains that several radiocarbon dates on charcoal and wood are at least 1,000 years older than Clovis sites, and suggests that occupation could have begun as early as 38,000 years ago.

And Brazilian archaeologist Niede Guidon claimed in a 1986 *Nature* paper that people were living at Pedra Furada in Brazil 37,000 years ago. Paul Bahn, writing later in *New Scientist*, said: 'Guidon's claims tend to be taken less seriously than those from the Chilean site [Monte Verde]. This is caused, in part, by her lack of publication ... but it is hard not to suspect that her gender and nationality may also be significant factors in this attitude.'

Michael Waters, of Texas A&M University, and Tom Stafford, who runs a private radiocarbon lab in Boulder, Colorado, recently scrutinised the record of Clovis sites in North America directly dated by radiocarbon analysis. The results from 11 sites with the most reliable ages suggested that Clovis technology lasted just 450 years, between 13,250 and 12,800 years ago, during which time the people making the projectile points spread across North America. The scientists got an even narrower range, of just 200 years, when they considered only the youngest age for the oldest Clovis site and the oldest age for the youngest Clovis site. This time span might turn out to be correct, but the statistics are questionable and calibration of the ages remains in dispute.

The appearance of Clovis technology throughout North America about 13,000 years ago fits two hypotheses. First, a rapid dispersal of Clovis populations across an American *terra nullius*. This is supported by demographic models of people travelling 13,000 years ago between the vast Laurentide ice sheet and the smaller Cordilleran ice sheet along the west coast. Perhaps the newcomers left the biface and blade technology, dubbed the Nenana culture, in transit. The assemblage was in Alaska about 300 years before the oldest Clovis sites considered in the Waters–Stafford study.

An alternative hypothesis is that the Clovis technology was introduced by a small group of migrants who spread rapidly through undefined populations already occupying the continent.

Meanwhile, genetics analysis suggests that the last leg of the trip, from Beringia to the main American continent, was between 11,000 and 16,000 years ago. It followed a dispersal from Central Asia between 20,000 and 25,000 years ago.

A recent review of archaeological and genetic evidence suggests that the colonists first took the coastal route, which was already open during the early phase of the deglaciation, before 15,000 years ago, with a second wave of migration possible via the inland ice-free corridor several millenia later. Recent DNA profiling on the 4,000-year-old hair of a palaeo-Eskimo from Greenland suggests that populations of Siberian–Beringian origin also spread to the extreme north.

About the time the Clovis culture flourished, 135 species, mostly of large mammals, perished. The animals had survived many previous climatic fluctuations and some traced their ancestry back to the start of the Quaternary, when North and South America were first joined. Climatic changes marking the final phase of the Pleistocene around 13,000 years ago were less severe than earlier ones. What caused the American megafauna's demise?

Towards the end of the last Ice Age, big-game hunters in Siberia crossed the Bering landbridge into Alaska. The human front advanced through the ice-free corridor to more hospitable lands to the south. They wiped out the megafauna in their path, perhaps within a thousand years of the incursion. That's the view of the University of Arizona's Paul Martin, the originator of the blitzkrieg extinction model, whose now famous paper 'The discovery of America' ignited a debate that would last for decades after its publication in *Science* in 1973. 'I propose that they spread southward explosively, briefly attaining a density sufficiently large to overkill much of their prey,' he wrote, '... unless one insists on believing that Palaeolithic invaders lost enthusiasm for the hunt and rapidly became vegetarians by choice ... or that they knew and practised a sophisticated, sustained yield harvest of their prey, one would have no difficulty in predicting the swift extermination of the more conspicuous native American large mammals.'

In Martin's scenario, 'naive' animals were wiped out soon after humans arrived in a region because they had not evolved defences against the new predators. There is strong evidence for blitzkrieg on islands, but some argue that islands are a special case, and results from them cannot be extrapolated to continents. It depends on the definition of an island, however. Australia has been a large island throughout the Quaternary, and America a much larger one during interglacials, such as the Holocene.

In their literature survey, Anthony Barnosky and colleagues say the naivety argument does not explain why some megafauna extinctions happened in Africa, which lost a few species, including some elephants and bovids. They argue that the timing of extinctions in Eurasia also weakens the blitzkrieg case, while other scientists counter that co-evolution with pre-modern hominins helped the megafauna wise up.

One of the earliest attempts at dating the American mass extinction pushed the radiocarbon method forward. The work was on bones from the disused Rancho La Brea tar pits—now a park in downtown Los Angeles—first excavated early last century. Hundreds of thousands of bones—from animals including the extinct dire wolf and sabre-toothed cat—have been recovered from the site, actually asphalt seeps. Some of them had what looked like cut-marks. They were impregnated with asphalt, and this posed serious contamination problems for the radiocarbon dating chemists, as asphalt is fossil carbon.

Ralph Wyckoff and colleagues, of the University of Arizona, Tucson, used an electron microscope to confirm that the bones retained the typical braided triple-helix structure of collagen. The scientists subjected the samples to rough treatment, including a wash in an organic solvent to get rid of the asphalt contaminants. It was a radical departure from Willard Libby's 'first do no harm' edict to quarantine samples from carbon, but Wyckoff's group had come up with a way to separate the sample carbon from that in the chemical cocktail. The scientists cooked the mixture in hot, strong hydrochloric acid to break the protein into its constituent amino acids, which they then separated on an ion-exchange resin like those in household water filters.

American dating expert Tong-Yun (Tommy) Ho and colleagues extended the amino acid composition studies. His team declared that the ultimate radiocarbon dates on bones would be on the amino acid hydroxyproline, which makes up 10 per cent of the total amino acid content in collagen but is rare elsewhere. Hydroxyproline and its precursor, proline, are more chemically resistant than the rest of the 20 amino acids in proteins.

When particle accelerators dedicated to AMS radiocarbon analysis came on line in the early 1980s, an undeclared race ensued between teams at the Oxford and Tucson facilities as both groups chased hydroxyproline. Gillespie and colleagues at Oxford got there first, but Tom Stafford and colleagues at Tucson extended the method.

A powerful new tool, compound-specific radiocarbon dating, had been forged, and is now used routinely. And the date on the last of the Rancholabrean fauna? Ho got an age of about 13,000 years.

The American megafauna bones are young enough to be dated directly by radiocarbon analysis, generally a more precise method than OSL and AAR. Figures are quoted with uncertainties of less than 100 years, compared with thousands of years for the Australian megafauna. However, what the American dates gain over the Australian ones in precision, they can lose in accuracy due to problems in calibrating the radiocarbon clock around the time of the Younger Dryas, which falls in the critical period.

In America, as in Australia, politics intrudes. Native American scholar, the late Vine Deloria, said in his book *Red Earth, White Lies* that 'American Indians, as a general rule, have aggressively opposed the Bering Strait migration doctrine because it does not reflect any of the memories or traditions passed down by the ancestors over many generations', and that 'advocating the extinction theory is a good way to support continued despoliation of the environment by suggesting that at no time were human beings careful of the lands upon which they lived'. The comments resonate with those of Australian archaeologist David Horton. In his book *The Pure State of Nature*, he said, 'if humans caused extinctions, then we needn't look at the present Australian environment as a natural system to be conserved'.

Research in the Americas is proceeding, however. Some scientists have been reading the droppings of an extinct species, while others have been reading the bones of an extant one. Still others have been seeking signs of an extraterrestrial cause of the extinction disaster.

10 Blast from the past

Almost 13,400 years ago, a big ground sloth visiting a limestone cave in Nevada left a calling card—a huge dung ball. The animal, possibly in the cave to give birth, was a *Nothrotheriops shastense*, one of about 30 species of ground sloth that went extinct in the Americas in the late Pleistocene. Weighing about 500 kilograms, the Shasta sloths were not as spectacular as their giant relatives that weighed up to 3 tonnes, but they were huge compared with the five species of sloths, all arboreal and under 10 kilograms, living today in tropical forests. And while not as charismatic as the sabre-toothed cats, horses, camels, mammoths and short-faced bear, they are among the most closely studied of the 40 genera of extinct American megafauna, thanks to information packed in their ubiquitous dung balls.

With front toes that turned under, the ground sloths were clumsy and slow-moving, relying on their diabolical front claws, their size and bony plates beneath their skin to protect them from the many predators in the Americas. The giant ground sloth stretched about 4 metres when it stood up on its hind legs to browse, while the Shasta sloths had a reach of 2.5 metres. Like their living cousins, the anteaters and armadillos, they had brain capacities that were small compared with their skulls. They evolved in South America, venturing north only after closure of the Isthmus of Panama 3 to 3.5 million years ago.

In arid environments like Nevada's Gypsum Cave, sloth dung balls dry out and harden, persisting for tens of thousands of years and retaining pollen, which gives insights into the animals' lifestyle and environment. Since they are not fossilised, they lend themselves well to radiocarbon dating and DNA analysis. Measuring between 11 and 18 centimetres in diameter, the distinctive brown balls, politely referred to as coprolites, are 'very consistent in appearance'. David Steadman, a palaeontologist at the University of Florida, said in an interview that if you called someone in America a sloth dung ball—something he didn't recommend doing—'there's an immediate three-dimensional image that everyone would have'. He is one of the few people on the planet who has smelled the dung of an extinct animal, but he's modest about the distinction, saying he stood on the shoulders of Paul Martin and other colleagues to earn it. The blast from the past happens when he resurrects the dung ball with water before sieving it to retrieve the plant material.

DNA analysis on Shasta sloth dung from Gypsum Cave by a team led by Hendrik Poinar shows that the animals had a more recent common ancestor with one living arboreal sloth than with another, suggesting that the forest lifestyle of living sloths originated twice in a case of convergent evolution. Steadman says the furry creatures had no enamel on their cheek teeth, so they couldn't handle phytoliths— the sharp silica crystals in grass—but they ate just about all the leaves on offer.

About 100 samples of sloth dung from hyper-arid south-west American caves have been radiocarbon dated. In 2005, Steadman and colleagues, including Paul Martin, compared 'last appearance dates' for the Shasta sloths in North America with those for other species of ground sloth in South America and for ground and tree dwelling sloths in the West Indies. The radiocarbon dates, reported in *Proceedings of the National Academy of Sciences*, were on dung balls and bones from continental America and on bones from Cuba and Hispaniola. The scientists wanted to find out whether the animals' exit coincided with climate change as the last Ice Age ended, or with the arrival of people in these regions.

The sloths disappeared 13,000 years ago in North America, hanging on a bit longer in South America, and until 5,000 years ago in the West Indies. Their exit times implicated humans in the cataclysm, with the dates tracking the path of people as they arrived in North

America via the Bering landbridge, penetrating South America within 1,000 years, and finally the West Indies in the mid-Holocene.

Had global climate change been to blame, the West Indian sloths would have expired in concert with their continental counterparts, the scientists argued. And the dung ball analysis showed that the weather should not have stopped the sloths making a living in their old haunts. The same plants grow there now as in their day, so destruction of their food source through climate change was not an option.

Research on ancient DNA from a surviving species tells a different story, however.

11 Bison

To get at the gold below, miners blast the frozen surface off a hillside near Dawson in the Yukon territory of Canada. To scientists watching on, there is something more precious beneath the surface of the permafrost—megafauna bones that haven't seen the light of day for tens of thousands of years. The bones jut out of chunks of icy mud that crack off the hill. One chunk gets airborne and demolishes a pine tree as the international team of scientists, including ancient DNA specialist Alan Cooper, now of Adelaide University, shelters beneath a bulldozer canopy. Other blocks tumble into a stream, and the scientists fish them out.

It's a ritual repeated each year in the long days of the sub-Arctic summer when temperatures, which get down to minus 50 degrees Celsius in winter, soar into the twenties. The miners, once perplexed by the brainy academics with a fascination for old bones, have grown accustomed to them, and now even let them have a go with their water cannon, used for the gentler excavation jobs. Some have caught the palaeontology bug.

The permafrost can preserve DNA for hundreds of thousands of years, and the scientists have been mining it for data on the deep past, focusing on Beringia. The team has recovered traces of plant DNA between 300,000 and 400,000 years old, and of bison, mammoth and horse DNA from Beringian soil. It also turned up a creature, closely related to the musk ox, not yet found in the fossil record.

In later work, DNA from fossils from the Yukon site was central to a big study, published in *Science*, aimed at piecing together the bison family tree and tracking demographic changes. The scientists dated and extracted DNA from 190 bison fossils from the Yukon, Alaska, Siberia, China and central North America. They also mounted an expedition to the supermarket to get samples from the two surviving American sub-species, the plains and the wood bison, which stand up to 2 metres tall and weigh up to 900 kilograms. They were working at the Henry Wellcome Ancient Biomolecules Centre at Oxford University, one of the few dedicated ancient DNA facilities in the world. The most famous is Svante Pääbo's laboratory at the Max Planck Institute in Leipzig, but the Oxford centre, which has extracted DNA from celebrities including Vikings and ancient hominins, has had its share of media attention. Cooper, who speaks with the candour once the norm in science, has now set up a rival lab—the Australian Centre for Ancient DNA (ACAD)—at Adelaide University. Like most scientists, he sees the media as a mixed blessing, and once put a futuristic-looking sandwich toaster in shot when a BBC crew was filming the Oxford lab's hi-tech sequencing equipment.

Ancient DNA facilities are designed to prevent contamination of the tiny samples with genetic material from researchers and other samples. Contamination has been the bane of the field since its advent in the early 1980s with the sequencing of part of the genome of the zebra-like quagga, extinct since the nineteenth century. The facilities are isolated from other genetics laboratories and have positive pressure gradients to keep dust and pollen out. Staff follow cleanliness protocol comparable to those in hospital quarantine wards. And supplies brought into the labs are cooked in ultraviolet light to destroy the DNA from sources including the urine of warehouse rats. Cooper notes that it also impresses visitors, 'particularly people who have been watching CSI, who have come to expect us to work with blue lights in the dark'. In their 'clean' room, Cooper and colleague Beth Shapiro donned body suits, surgical masks and gloves to extract DNA from half-gram bone samples.

They wanted to isolate DNA from the mitochondria, the tiny structures outside cell nuclei that power cells. These metabolic power-packs started out as bacteria in a symbiotic relationship with other primitive lifeforms more than a billion years ago, trading energy for real estate. They have a separate, circular, genome carrying the instructions for the complex series of reactions that convert food into energy. Mitochondrial DNA (mtDNA) is maternally inherited, in contrast to most nuclear DNA, in which maternal and paternal genes are shuffled in a process called recombination. (An exception is nuclear Y-chromosomal DNA, which is passed down the paternal line.)

As with nuclear DNA, the mitochondrial code is written in the letters A, C, G and T, denoting the chemical bases, or nucleotides, adenine, cytosine, guanine and thymine, arranged on each strand of the double helix DNA molecule along a backbone of phosphate and the sugar deoxyribose. Each base on one strand is joined by a weak hydrogen bond to a complementary base on the other to form 'base pairs'. A always bonds to T, and G to C. Mitochondrial DNA mutates more rapidly than nuclear DNA, so it gives a strong signal to geneticists using it to compare genetic variation between individuals or populations or through time.

DNA can survive for more than 50,000 years in bone and teeth. Dung is also a popular source among geneticists because it betrays the DNA of the animal—from its intestinal cells—along with that of its food and parasites. And animals like the ground sloth, which had communal toilets, raise the possibility of sampling 'whole populations', Cooper says enthusiastically. Hair, sediments and glacial ice are other sources. However, DNA degrades, especially in warm, humid environments, mainly through hydrolysis, a chemical reaction with water, or through oxidation, and ancient DNA experts have only fragments to work with. Biochemical calculations show that DNA dissolved in water will degrade so rapidly that no intact molecules will be left after 5,000 years. Due to the higher copy number of mitochondrial genomes per cell, the chances of mitochondrial genes surviving are better than those of nuclear genes, so most work on ancient DNA has centred on mitochondria.

The scientists amplified bison mtDNA using the polymerase chain reaction (PCR), a method developed by the American chemist Kary Mullis in the early 1980s to make multiple copies of stretches of DNA, often particular genes. Before sequencing it, they cloned it, standard procedure to identify contamination from people handling the material and a kind of post-mortem DNA damage called miscoding lesions, in which one base gets substituted for another, mimicking mutations and confounding the phylogenetics.

By comparing the DNA hoofprints of bison of the same period from the various regions, Cooper and colleagues pieced together branches of the family tree and assembled them into a Pleistocene genealogy, unravelling details of the beast's tangled past. They worked out when the branches diverged by calibrating the bison molecular clock directly from the bones, using radiocarbon dates and DNA sequences to see how quickly mutations accumulated in the genome. The method, only made possible when scientists learned how to get DNA out of ancient bones, departs from the early molecular clock calculations and is more accurate. The earlier work involved counting the number of mutations in modern DNA of two closely related species that had diverged from a common ancestor at a time estimated from a sketchy fossil record. This method was plagued by uncertainty in the taxonomy of the common ancestor and in dating it.

A study by a team led by Oxford University's Simon Ho showing that the mitochondrial DNA mutation rate was not constant set the alarm bells ringing about molecular clock calculations based on modern DNA. The team showed that the rate of evolution appeared to speed up towards the modern day. The implication is that molecular clock calculations are out, particularly those involving the recent past.

The team studying bison DNA found that the code of ACGT bases changed at a rate of 32 per cent per million years, and that the most recent common maternal ancestor of all the animals in the study lived about 136,000 years ago in North America. Genes flowed between Asia and North America via the Beringian landbridge until the ice sheets isolated northern and southern populations during the last Ice Age. The groups interbred after the ice sheets began to retreat about 16,000 years ago, but were separated again by rising sea level at the end· of the Pleistocene 12,000 years ago. Spruce forests flourished across Alberta, and peatland in western and north-western Canada erected ecological barriers. At the same time, trees and shrubs invaded the Beringian grasslands that had sustained the bison.

Big populations rack up more DNA mutations than small ones, so a group's genetic diversity is related to its population size. Sophisticated computer models, also used to study ecology and to track virus evolution in epidemiology, are needed to glean details on population dynamics and sizes from the genetic data. The mathematics behind the models is daunting. 'What do you do?' is not the sort of question one should ask modelling gurus at parties—you risk a monologue on why Bayesian skyline plots are better than ordinary skyline plots for identifying the most appropriate demographic model to plug into a coalescent model. The bison number-crunchers cracked a problem besetting all ancient DNA work—how to get a handle on whether the ancient population was growing, declining, or steady.

The team ran the new models on DNA sequences from the modern bison lineage alone to see if the results squared with historical facts. The models correctly identified a severe population bottleneck in the late 1800s, when everyone used the Winchester repeating rifle. Applied to all of the bison DNA sequences, the modelling suggested a boom—bust population scenario in the past, driven by climate. The Beringian population reached a peak of several million animals in Oxygen Isotope Stage 3 (60,000 to 30,000 years ago) before going into freefall just as the planet headed towards the coldest phase of the last Ice Age, thousands of years before the Clovis big-game hunters came on the scene. The population passed through the smallest bottleneck around 11,600 years ago, soon after the arrival of people, getting down to about 10,000 individuals. It quickly rebounded but never came near the levels of Pleistocene park.

The geneticists had come up with the first hard evidence for a climatic impact on megafauna. The findings were later backed up by palaeontological research on mammoth and horses.

If other species had gone the same way as bison, perhaps climatic factors had worn them down so much they could not handle the stress of the wily new predator. Similar arguments have been advanced for Australia, and, in Cooper's view, the debate has shifted from discriminating between climatic and human scenarios to assessing the relative contribution of each. At the time of publication, Cooper, and ACAD colleague Jeremy Austin, renowned for his work showing that DNA does not survive in ancient insects preserved in amber, were about to start similar studies on the other continents, including Australia.

The bones lodged in the permafrost are the last traces of the native Beringian bison, for the animals went extinct in the region a few hundred years ago and left no direct descendants. However, each summer, a dull thumping is heard in Yellowstone National Park. It is the sound of bison head butting as males vie to brain their rivals and pass on their genes. Like most big animals, bison reproduce slowly, and the calves will not be born for nine months. Bison are now protected on reserves in numbers high enough to secure their future. Others are on ranches servicing game hunters or supplying the game meat market with lean protein favoured by the heart-smart.

However, the beasts still get a rough time. Oil companies and conservationists in Alaska have locked horns over a campaign to reintroduce the species there. Some politicians oppose the move, and the government has given orders to shoot animals straying across the border from Canada. The genetics research refutes one of the main arguments against the reintroduction—that the original Alaskan bison was a different species from the modern population. A comparison of genetic data on 15 groups—classified by palaeontologists as distinct species on the basis of bone morphology—failed to turn up evidence that the groups had stopped mating. Once thought to be of recent Beringian stock, modern bison descended from herds living south of the ice before the last Ice Age, according to the ancient DNA study. Their lineage diverged from the Beringians' 64,000 to 83,000 years ago. The palaeontologists had been extreme 'splitters' in their taxonomic schemes, underestimating the amount of individual variation within a species through time and space. The geneticists, meanwhile, are 'lumpers', and Cooper's group has also challenged the number of species assigned to horses and lions.

So how did the bison survive when most of their big contemporaries died out? Although bison genetic diversity plummeted, it did so off a high base. The research revealed high mitochondrial diversity in ancient bison, reflecting the big populations and confirming what the animal liberationists always knew—that each bison was an individual. At least some of this diversity is likely to have been mirrored in the

nuclear genome, which codes for physical and behavioural characteristics. In fact, it is the individual variation within the species that has
complicated the taxonomy and preoccupied palaeontologists in heated
rows over it for years. Perhaps the shaggy brown bearded beasts had
enough individuals with the right stuff—mobility, for example—to
withstand repeated shocks, but whatever sent bison numbers up and
down might have wiped out the other big animals.

Some American researchers contend that the Younger Dryas, named
for the flower *Dryas octopetala* that grows in the Arctic tundra, was
to blame. The big chill hit amid the warming period after the last Ice
Age. It was recognised in the 1930s and 1940s in ancient pollen in
varves laid down in Swedish lakes and bogs. Radiocarbon dating and
the number of annual layers in Greenland ice cores have enabled scientists to fix the Younger Dryas at 12,900 to 11,600 years ago. Scientists argue about its geographical extent, but its signature is heaviest
in continental regions around the North Atlantic. It lasted about a
millennium. Then, within two decades, temperatures soared.

Changes in ocean circulation caused the Younger Dryas, says ocean-
ographer Wally Broecker, of the Lamont-Doherty Earth Observatory
in New York. Glaciers started retreating about 14,000 years ago, their
meltwaters banking up in the interior of North America. Then 12,900
years ago, after further retreat of the eastern glaciers, the meltwater

Dryas octopetala

flowed east across the region that would become the Great Lakes, finally pouring into the Atlantic Ocean. An estimated 10,000 cubic kilometres of freshwater diluted the Atlantic, disrupting the ocean circulation and stopping the deep 'conveyor belt' current that delivers heat to northern regions. The situation persisted as long as the freshwater flowed. Today, the conveyor is again in motion and drives the Gulf Stream, a current of warmer water flowing north at a depth of about 800 metres. A torrent the size of the Amazon River, the Gulf Stream surfaces near Iceland, releasing heat before sinking under the weight of its salt. Europe would be much colder if not for the Gulf Stream. The movie *The Day after Tomorrow* drove home the risk of a Younger Dryas-like winter if human-induced global warming shut down the Gulf Stream.

Some scientists have a more dramatic explanation than Broeker's for the Younger Dryas.

Catastrophism, an old philosophy long out of favour but undergoing a resurgence, has been invoked.

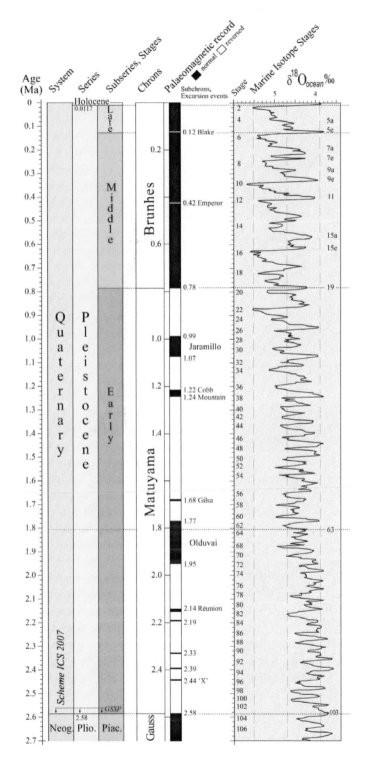

Palaeoclimatic and palaeomagnetic record during the Quaternary. (Adapted from
www.qpg.geog.cam.ac.uk/)

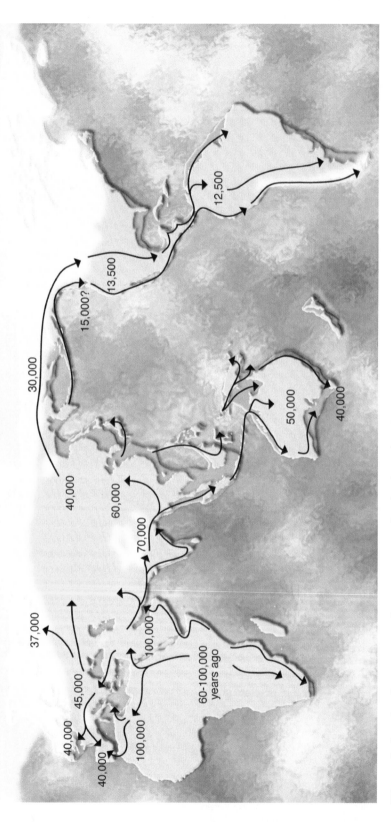

Dispersal of modern humans according to genetic and archeological evidence. Dates are in years before present. Coastlines during glacial times are shown in pink. (Adapted from Douglas Palmer, *The Origins of Man*, New Holland Publishers (UK), 2007)

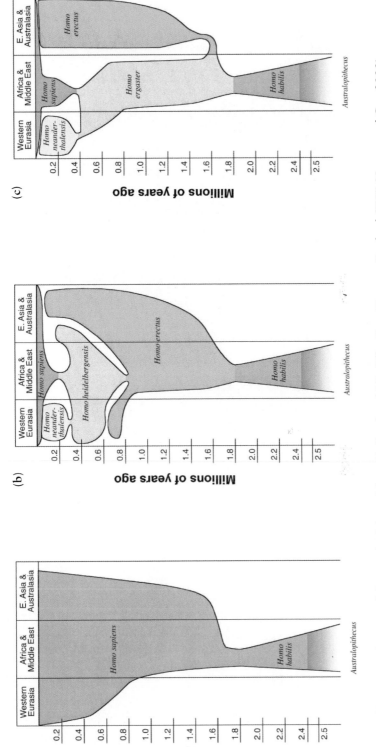

Models of human evolution and dispersal. (Adapted from R. Boyd and J.B. Silk, *How Humans Evolved*, W.W. Norton and Co., 2000)

(a) According to M. Wolpoff. There is only one human species and there were no speciation events since about 2 million years ago.

(b) According to G. Phillip Rightmire. The African *H. ergaster* and the Asian *H. erectus* should be classified as a single species.

(c) According to R. Klein. *H. ergaster* evolved in Africa 1.8 million years ago and soon spread to Asia, where it became a second species, *H. erectus*.

Claudio Tuniz at the archaeological site of Jwalapuram, in the Kurnool District of Andhra Pradesh, India, in 2007, looking at the 2.5-metre deposit of ash from the Toba super-eruption, 74,000 years ago. (Photo: C. Tuniz)

Claudio Tuniz and Rhys Jones at the AMS facility of the Australian Nuclear Science and Technology Organisation in 1993. (Photo: ANSTO)

Liang Bua cave, Flores, Indonesia. Sector VII, where the remains of *H. floresiensis* were found, is on the right-hand side. (Photo: Djuna Ivereigh/ARKENAS/indonesiawild.com)

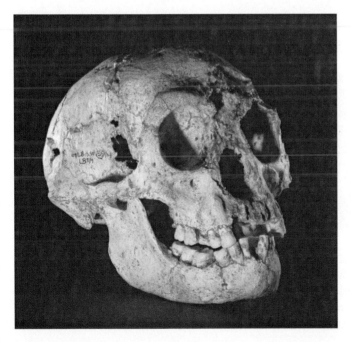

Homo floresiensis cranium. With a brain capacity between 380 and 420 cubic centimetres, the hominin status of the hobbit is still being debated. (Photo: Djuna Ivereigh/ARKENAS/ indonesiawild.com)

The Walls of China in Mungo National Park. (Photo: Michael Amendolia 2005 with permission of the traditional land owners)

Large-scale archaeological excavation directed by ANU archaeologist Wilfred Shawcross in 1976 at Lake Mungo, Willandra Lakes Region, a World Heritage listed cultural and environmental site in southwestern New South Wales. The skeletal remains of Mungo Lady were found nearby in 1968, and Mungo Man in 1974, both by geologist Jim Bowler. (Photo: John Magee)

The 41,000-year-old skull of a giant kangaroo, *Protemnodon anak*, the youngest known Australian megafauna specimen, from Mt Cripps, Tasmania. (Photo: Queen Victoria Museum and Art Gallery)

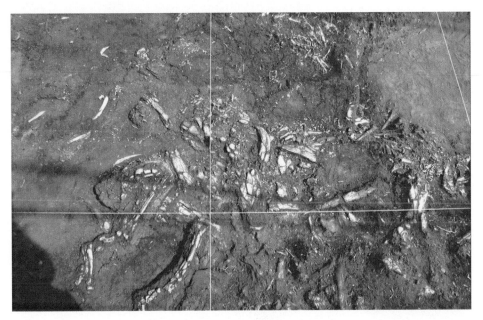

Megafauna excavation in 2004 at Black Creek Swamp on Kangaroo Island, South Australia, directed by Flinders University palaeontologist Rod Wells. Leg bones from the extant grey kangaroo are visible as well as teeth from the extinct herbivores *Diprotodon* and *Sthenurus* and the extinct carnivore *Thylacoleo*. (Photo: Richard Gillespie)

Reconstruction of *Homo neanderthalensis,* Neanderthal Museum, Mettmann, Germany. (Photo C. Tuniz)

Diprotodon optatum. The giant marsupials were among the casualties in the mass extinction of the Australian megafauna. (Illustration: Peter Murray)

The extinct flightless bird *Genyornis newtoni* dwarfed the emu. (Illustration: Peter Murray)

12 Cosmic impact

In the morning of June 30, 1908, reindeer herders asleep in their tents at a camp near the Podkamennaja Tunguska river in central Siberia were blown into the air by a big explosion while everything around them was covered in smoke from burning trees. At the same time, people in the town of Kirensk, 400 kilometres away, noticed a gigantic pillar of fire, followed by a large cloud of black smoke on the horizon. A train on the Trans-Siberian line, 500 kilometres away, was hit by burning debris falling from the sky, followed by black rain. Vibrations were recorded by seismograph stations around the planet and transoceanic ships had their communications blocked. Yet the event was barely noticed by the Russians, in the midst of dramatic social and economic unrest. Only in 1921, after the Russian Revolution, was a scientific expedition sent to study the Tunguska explosion. It found that 60 million trees had been knocked down over an area of 2,150 square kilometres. Most trees at ground zero were burnt. Many lie on the ground today, aligned in the same direction from the epicentre of the event.

In the 1990s, an Italian group analysed rings from Tunguska trees that survived the explosion. Metallic spherules in the resin had a composition suggesting a meteoritic explosion. Similar magnetite and glassy spherules were found in samples from soils and sediments. The power of the explosion has been estimated at 10 to 20 megatons, 1,000 times bigger than the Hiroshima bomb. No impact craters or meteorite fragments were found, so the nature of the hypothesised cosmic body remains uncertain.

Hypotheses on the cause of the 1908 event include alien intervention, nuclear explosions and anti-matter effects, but most serious scholars believe the blast was caused by an extraterrestrial body that exploded about 10 kilometres above the Earth's surface. Some groups believe this is the typical fate of stony asteroids of several metres in radius entering the Earth's atmosphere at hypersonic speeds. Carbonaceous asteroids or comets would break up at a much higher altitude. Impacts of this magnitude by comets or meteorites occur every few thousand years.

Could a similar extraterrestrial strike be behind the demise of more than 30 genera of large-bodied mammals in the Americas at the end of the Pleistocene? Some scientists say an extraterrestrial impact over northern America around 12,900 years ago triggered the Younger Dryas, ended the Clovis culture, and took out the megafauna. In conference presentations and a paper in the *Proceedings of the National Academy of Sciences* a team led, appropriately, by Richard Firestone, of the Lawrence Berkeley National Laboratory, has presented its evidence for the comet catastrophe hypothesis from sediments from sites associated with Clovis technology, the Younger Dryas, and megafauna.

The team reported that the sediments had a high concentration of extraterrestrial (ET) markers—glass-like beads, soot, the carbon nano-particles known as fullerenes or 'buckyballs', and the exotic element iridium, present at high levels in asteroids and meteorites but rare on Earth. These materials were absent from other layers of the stratigraphic record. The glassy carbon beads implied melting at more than 4,000 degrees Celsius. These materials are well known from other extraterrestrial impacts, such as the Cretaceous–Tertiary boundary hit that wiped out the dinosaurs, and the Tunguska explosion. Electron microscope analyses showed the glassy spherules were rich in micro-diamonds. Diamonds are formed deep in the Earth by the compression of carbon at a pressure of several gigapascals, conditions duplicated on the surface of the planet only by the impact of massive extraterrestrial bodies. Meanwhile, buckyballs trap traces of noble gases, such as helium and argon, in their 60-atom spherical cages. Noble gases from Earth have different isotopic ratios from those of extraterrestrial origin.

At Gainey, near Chicago, archaeologists have unearthed Clovis-style arrow and spear points in 12,900-year-old sediments. The site bears the highest concentration of ET markers, suggesting that Gainey was near the epicentre of the explosion. ET markers have also turned up at famous Clovis sites including Murray Springs, Arizona, and Blackwater Draw, New Mexico. They appear in many Clovis and megafauna sites at the base of a 12,900-year-old layer described by the University of Arizona's Vance Haynes as a 'black mat'.

Cosmic catastrophists argue that an extraterrestrial body exploded over northern America, probably around Ontario or Hudson Bay: ET indicators are in the highest concentration in the north of the continent, tapering off towards the south. The absence of craters suggests the body was a comet, made of ice but with some carbon and rocks, at least 5 kilometres in diameter. It probably exploded on approach to the planet, like the comet Shoemaker–Levy in its 1994 encounter with Jupiter. According to Firestone's team, the heat of the explosion accelerated the melting of the Laurentide ice sheet, which was under way at the end of the last Ice Age. The fallout darkened the ice, decreasing the amount of heat it reflected into space, and reinforcing the melting. The high-energy shock wave fractured the ice sheet, which extended south to present-day Michigan at the end of the Pleistocene. It opened channels that dumped water eastward into the North Atlantic, blocking the oceanic conveyor belt and sparking the Younger Dryas. Wildfires swept across northern North America, causing the megafauna mass extinction, the Firestone team contends.

Many scientists disagree, arguing that Wally Broecker's oceanographic circulation model for the Younger Dryas is more parsimonious. Some call for sampling of a broader range of sediments for ET markers. They say that the soot layer attributed to the comet hit is conspicuously absent from the well-dated Greenland ice sheet. Other experts caution against reading too much into exotic spherules. Similar material of cosmic origin, rich in iridium and other rare elements, falls continuously from the upper atmosphere.

* * *

Paul Martin, in his 2005 book *Twilight of the Mammoths*, has a suggestion for resolving the American colonisation and megafauna extinction arguments:

> Australian archaeologists have in the last four decades radically extended their chronology of human arrival to or beyond the limit of radiocarbon dating at dozens of sites, while archaeologists hot on the trail of pre-Clovis colonisation have failed to nail down any robust evidence of North American sites that is acceptable to the community of archaeologists as a whole. The discrepancy should trigger serious revisionary thinking. Perhaps American archaeologists in search of pre-Clovis sites need to hire some Australians. Aussies seem to be capable of finding and agreeing on the existence of sites tens of thousands of years older than the late-glacial fluted points and fishtail points that are the oldest artefacts unclouded by controversy in the Americas.

But Martin—who marked and passed the PhD thesis of one of the main opponents of his blitzkrieg hypothesis, Judith Field—has not visited Australia for two decades, and the political landscape has changed. Tim Flannery, with his man-the-hunter stance, is seen as fair game. Like Paul Martin, he is targeted in startling attacks in the media and in wordy treatises in minor and semi-popular journals.

13 Cool science, hot politics

'The Australia of Flannery's prehistory is populated by blokes and its destiny determined by the barbies [barbeques] they had when they first arrived. Once the hangover wore off, they stoically made the best of the mess they had created,' wrote Wollongong University's Lesley Head in a paper in the literary journal *Meanjin*. Tim Flannery, who has written 27 books, held the coveted Chair of Australian Studies at Harvard University, and was Australian of the Year in 2007, cops this sort of critique in his homeland. Many of his colleagues do, too. Metadata analysis suggesting human agency in extinctions is dismissed as 'the big bloke theory', and the blitzkrieg hypothesis is attacked on the grounds that it is 'macho' and belittles women's food-providing role. 'The mythology, among both Aboriginal and non-Aboriginal people, is that Aborigines are big-game hunters,' David Horton wrote in his book *The Pure State of Nature*. 'It is a male mythology, as such mythologies usually are, all over the world ... In fact, families were largely fed by women, and largely fed on small items of food.'

In a politicised prehistory, the perceived or imagined political implications of research are viewed as a legitimate concern of scientists. And while most of the salvos are fired at the blitzkrieg hypothesis, research supporting indirect human impacts is targeted, too.

John Benson, a plant ecologist at Sydney's Royal Botanic Gardens, has claimed that Flannery's views on firestick farming have been seized upon by farmers bent on land clearing. '... simplistic statements made by reputable authors such as Tim Flannery can be used to justify ongoing damage to the Australian environment,' Benson wrote in a letter to *Quarterly Essay*, an Australian magazine of the 'left'. In a reply, Flannery pointed out that, as a member of the Wentworth Group of Concerned Scientists, he was one of the architects of sweeping reform to land clearing legislation in New South Wales. He challenged Benson to do more science and cut the 'cheap shots and polemic'.

Stephen Wroe, Judith Field and Richard Fullagar wrote in the Australian palaeontology journal *Alcheringa*:

> Interpretations of megafaunal extinction have been manipulated by politicians and special interest groups. It has been argued that because it is now 'known' that Australian Aborigines rapidly wiped out the megafauna, pastoralists are simply applying replacement therapy by stocking hard-hoofed cattle. Others have used this 'fact' to discredit Aborigines as environmental custodians.

Many critics fail to mention the part of Flannery's thesis that has Aborigines reaching a post-extinction equilibrium, protecting the biota from further losses until European colonisation. And Flannery is still waiting to see examples of negative impacts of his work. On the Magee–Miller monsoon and extinction study, Wroe, who describes himself as a fence-sitter, was quoted by the ABC as saying: 'It's a circumstantial case that can be seriously misused by special interest groups in society, from environmentalists to people opposed to Aboriginal control of land.'

Then there is the charge that arguing for human impacts hangs an albatross around Aborigines' necks. 'Many archaeologists are sensitive to the fact that placing the blame for prehistoric extinctions entirely on the ancestors of surviving indigenous people can fuel vilification,' Wroe said in an editorial in *Quaternary Australasia*. One scientist, who didn't want to be named, pointed out that there was nothing in the data to suggest intent. 'All the evidence is saying is that these people were probably not fundamentally different from all other humans on the planet', he said in an interview. 'The main preoccupation is providing shelter and food for their families. To take it any other way, I think, is racist. Very localised activities happened to have dramatic and highly unanticipated consequences.'

Wroe blames the media for fuelling the controversy: 'The blame game has long been a hallmark of the megafaunal extinction debate, with findings typically presented as strong support for human or climatic causation, a tendency aggravated by media that thrive on simplicity and conflict'. However, he was quoted by *The Sydney Morning Herald* as saying of Flannery: 'Just because a guy is well known does not mean that he knows what he is talking about. I've got a fairly cynical view of Tim. He's an opportunist.' Judith Field was quoted in the same article as saying, 'Tim doesn't let the facts get in the way of a good story.' Of the scientist who has published more than 120 refereed papers, on everything from the description of new marsupial species from New Guinea to the evolution of maximal body size, she reportedly said: 'He does a lot of broadbrush stuff, with broad consequences, and some of it is just plain wrong … Most of our hypotheses are tested with facts, and that underlies the work we do. But most of what Tim does is conceptually driven, and not based on data. And he has been selective in his use of data.'

So close are the Australian and American politics of extinction that the arguments could have been scripted on either continent.

The *Mammoth Trumpet*, an American magazine about people and megafauna, reported on a 'Clovis and Beyond' conference held in 1999 at Santa Fe, New Mexico, which could have been mistaken for a dress rehearsal for the 2001 National Museum of Australia forum and the Quaternary Extinctions Symposium, held in Naracoorte, South Australia, in 2005.

Some of the 1,400 delegates in Santa Fe argued for a long human–megafauna overlap, exonerating people for the beasts' destruction. They charged that researchers had not dug in the right locations or looked hard enough for pre-Clovis archaeology. The University of Arizona's Vance Haynes, a long-time 'Clovis First' advocate, parried: 'What archaeologist worth his or her salt would refrain from excavating below a Clovis level because of a preconception that there would be no archaeology there? The concept is certainly unscientific.' In his 1964 *Science* paper, 'Fluted projectile points: Their age and dispersion', Haynes had assessed the distribution and ages of sites with Clovis and other spear points.

University of Alberta archaeologist Alan Bryan told the Clovis and Beyond conference that the evidence from the Monte Verde and other early sites in South America meant that the 'Clovis First model has been disproven and that there must be earlier sites in North America'. Haynes countered, 'If Monte Verde is a valid 15,000-year-old site, this fact does not mean that sites previously rejected for lack of compelling evidence have any more validity now than they did under earlier scrutiny. Nor does it mean that equivocal pre-Clovis sites are now less so.'

Joe Watkins, of the Federal Bureau of Indian Affairs, summed the atmosphere of the conference up. 'As has been pointed out to me by most American Indians who are aware of this conference, I am the only American Indian speaker here and I am suspect because I'm also an archaeologist. We, the archaeological and the American Indian communities, are tied tail-to-tail like two wild cats, fighting and spitting while attempting to inflict damage on each other.'

Paul Martin's nemesis in academe is archaeologist Donald Grayson, of the University of Washington. In a heated exchange in the usually cool scientific literature, Grayson and David Meltzer said in a 2003 paper titled 'A requiem for North American overkill': 'Martin's position has become a faith-based policy statement rather than a scientific statement about the past, an overkill credo rather than an overkill hypothesis.' Stuart Fiedel and Gary Haynes hit back in a paper called 'A premature burial: Comments on Grayson and Meltzer's requiem for overkill', accusing the authors of 'theatrical posturing', and adding that 'Although their critical assessment of the Late Pleistocene archaeological record is laudable, Grayson and Meltzer unfortunately make numerous mistakes, indulge in unwarranted *ad hominem* rhetoric, and thus grossly misrepresent the overkill debate.'

Running in counterpoint to political arguments are irrelevant ones. One centres on the lack of Australian 'kill sites' and a paucity of American ones—just 14. 'For the first Australians, there is no smoking gun in the form of murdered megafauna, there are no specialised weapons, and their immediate ancestors were almost certainly not systematic hunters of big animals,' Wroe, Field and Fullagar write in *Nature Australia*, a magazine previously produced by the Australian

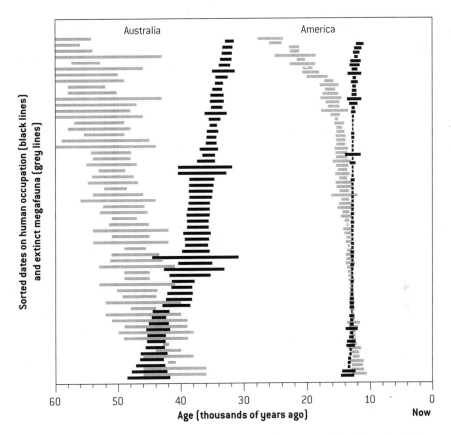

Dates on human occupation of Australia and North America (black lines) and dates on extinct megafauna (grey lines), where the length of lines represents uncertainty in the dates. Megafauna dates from both continents extend much further back in time than the first human arrivals in Australia about 46,000 years ago and in America about 13,000 years ago.

Museum. They admit that 'we know almost nothing about the culture of first Australians', but then go on to say, 'if they were anything like all known societies from low latitudes, then their diet comprised around 70 per cent vegetable matter and most meat eaten was in the form of small game'. Their American counterpart, Grayson, says: 'Where's the spear point sticking out of a camel or ground sloth? Clovis people absolutely did not chase these now-extinct animals relentlessly across the North American landscape.'

Martin argues that a blitzkrieg would wipe out prey so quickly there would be no time for the evidence to be buried and locked in the stratigraphic record, and Flannery points to the bone preservation problem in Australia. There are moves to look at the question statistically, estimating the number of kill sites that would be expected to show up if the giant animals were wiped out in a hunting frenzy. Todd Surovell and Nicole Waguespack, of the University of Wyoming, in a paper titled 'How many elephant kills are 14? Clovis mammoth and mastodon kills in context', argue that the 14 known elephant kill sites in North America are a lot, and '... in comparison to the Old World record, Clovis peoples seem to have exploited elephants with much greater frequency than in any other time and place'. The question is academic anyway. Sites with well-preserved bones, such as the huge moa butchering sites in New Zealand, colonised about 750 years ago, leave little scope for plausible deniability. But the kill site question will not resolve the debate over continental extinctions. The discovery of one such site in Australia would prove only that the first Australians killed one animal.

Another facile argument centres on the absence from the early Australian record of big-game hunting technology like the Clovis spear points. Blitzkrieg supporters counter that people could put a big dent in an animal with a sharpened stick, or wipe out a species by targeting juveniles.

Flannery, whose detractors in 'progressive' circles are joined by the rabid right (the neoconservative blogosphere attacks him for believing in human-caused global warming), is bewildered by how personal the attacks can get. 'I've been accused of everything from rape to racism,' he says. 'Everyone has a construction in their head of everyone else and sometimes there's not very much you can do to influence it, particularly when you've got a public persona. I'm not saying they [Aborigines] did it but we didn't. I'm saying it's basic human nature.'[1]

Martin, for his part, appears to be growing weary with the bickering. 'We have seen the genocide of 20 million people. You and I and our ancestors are not immune to behaviours which culturally we abhor. We are a species which is capable of quite a lot of killing. Should I not consider at least the possibility that our ancestors entered a landmass new to them and found animals that were absolutely naive to respond to this two-legged predator? We assume, "Oh no! We couldn't possibly do it."'[2]

Vance Haynes, who was named Geoarchaeologist of the Twentieth Century at the 1999 Santa Fe meeting, proposed: 'In the future, the scientific investigations of all potential pre-Clovis sites must include on-site evaluation of evidence as it is recovered. The standards today should be no less than they were with Folsom in 1927. It is apparent, as always, that available evidence and interpretation of data becomes more or less subjective depending on the bias of the interpreter.'

14 Extinction science

The little South Australian town of Naracoorte was proud to be hosting the Quaternary Extinctions Symposium in 2005. Organisers had promised that the event, the first in eight years, would attract 'eminent international extinction scientists' to the Limestone Coast town, 340 kilometres south-east of Adelaide.

Conference sessions were held in the historic town hall, and there was a guided tour of the Naracoorte Caves. To liken the World Heritage listed caves to a cathedral would be more of a compliment to Christopher Wren than to the sculpted limestone labyrinth, which awakens an atavistic urge to move in. Flinders University's Rod Wells discovered the fossils of extinct megafauna in the caves in 1964, and scientists have been studying Naracoorte's rich deposits ever since.

Just about everyone who is anyone in the great megafauna extinction debate was at the symposium—palaeontologists, geologists, physicists, biologists, geochemists, palaeoclimatologists, geneticists and archaeologists. No other conferences attract such academic diversity, with delegates drawn from fields ranging from the hard sciences to the almost literary. At the official dinner before the event, delegates distributed themselves among tables according to discipline and faction. Local dignitaries reading well-worn speeches ensured delegates were thoroughly backgrounded on the virtues of the region, especially its environmentally sustainable wine and 'small seeds' industries. The nature of the latter taxed even the greatest minds present, and the scientists, on learning that the small seeds were from lucerne, would have to get their brains around the concept of an environmentally friendly introduced pasture.

When the busy schedule of the presentation of papers got under way the next day, session chairs graciously introduced speakers about whom they had little nice to say in private. The conference started like any other, with researchers reporting their results, but the friction between two world views soon ignited spot fires, which diplomatic convenors had to douse quickly. Scientists had their vocabularies expanded by researchers declaring, without any hint of irony, their intention to 'deconstruct' aspects of the extinction problem. They learnt a new scientific term—'partial articulation'. Applied to some of the Cuddie Springs remains, the term seemed a bit Zen only when one thought about it. They learnt that science's simple elegance—parsimony—was passé because it was simplistic: a mosaic of extinctions happened, for various reasons, at different rates, times and in different places. They learnt that the focus on dating and temporal association, rather than on archaeological context, 'sidetracks the whole issue and the whole debate'.

And they learnt a new methodology—science by democratic vote, with most archaeologists viewing the oldest dates for human colonisation 'very problematical'. This approach, a departure from the protocol of the scientific method's research, peer review and publication, would be developed the following year when Judith Field conducted a poll on Ausarch, an Australian email list for archaeologists, asking subscribers if they believed humans were mostly responsible for megafauna extinctions.

Delegates periodically galloped off on wild *Genyornis* chases for kill sites and hi-tech stone hunting tools. Untestable hypotheses were advanced: if the giant herbivores could handle themselves against a marsupial carnivore and a big lizard, surely they could handle human hunters! Conspiracy theories were formulated: the Roberts team had focused on articulated remains to eliminate 'younger numbers'.

And they were assured by Field that direct ESR dates on the Cuddie remains were 'imminent'. Delegates at least agreed that the corpus of extinction knowledge had grown since the previous symposium eight years before, but there was still much to learn. In the prevailing academic climate, with the 'publish or perish' culture replaced by the 'patent or perish' one, there was some agonising at the symposium over the relevance—and marketability—of work that merely illuminated our species' remote past. There was talk of the application of Pleistocene research to the environmental problems of the present—global warming, habitat destruction and the identification of endangered species most at risk.

An account of the South Australian Government's attempt to reintroduce a sub-species of the tammar wallaby damped any optimism, however. A colony of the sub-species, extinct in Australia since the 1920s, was found on New Zealand's Kawau Island, where the animals had gone feral after being introduced there more than 100 years ago. Australia got some back and launched a captive breeding program ahead of a release effort. But a fox took three of the first 10 wallabies set free, in 2004. The conference heard that the government had a fox baiting program in place but some graziers, angry about the reintroduction of a competitor for their stockfeed, had ditched their own programs. Still, the conference was held before the environment returned to the political agenda. A worsening water crisis and the unsettling fourth assessment report of the Intergovernmental Panel on Climate Change made global warming, at least, an election issue in 2007.

＊＊

Scientists may never find the smoking gun, or smoking firestick, that will enable them to discriminate between a blitzkrieg and a slow burn. Their arsenal of dating weapons may never be that powerful. And the role of regional climatic forces, like El Niño, may never be fully understood. Answers will come only from the best science—from palaeoclimatologists and geologists, from chemists doing stable isotope analysis of bone and teeth, geneticists working on ancient and modern DNA, ecologists refining their population dynamics models and archaeologists sorting out the cultural sequences. But the dating specialists will remain the big guns. Hopes are high for methods combining the best uranium-series, ESR and luminescence techniques. And the radiocarbon revolution is far from over; specific-molecule dating is a rapidly growing field, with moves to push the event horizon even further back in time, and sort out the calibration problems.

Scientists can, theoretically, also work out the sizes of founder human populations. Genetic variation within modern Maori, for example, suggests New Zealand was colonised by as few as 100 people. Researchers could also derive demographic information from modern and ancient human DNA in studies similar to the American bison research. However, problems in DNA preservation can throw up serious obstacles, and politics is always heated among that success-ful, and destructive, species—*Homo sapiens*.

That genome has been the battleground of supporters of the two competing theories for the evolution of modern humans. Ancient remains from Lake Mungo and Kow Swamp have been at the centre of the scientific controversy amid bitter political disputes over the repatriation of skeletons.

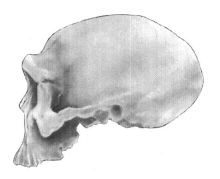

III
ORIGINS

15 Gene wars

'The destruction of Eve' is how one headline put it. Australian scientists, it seemed, had thrown a spanner in the works of human evolutionary theory. They claimed to have extracted DNA from the Mungo Man skeleton, which had just been redated to around 62,000 years old. It was, they said, the oldest DNA ever recovered from a human: the closest contender was a Croatian Neanderthal 20,000 years younger.

The team, which included multiregionalist Alan Thorne, said Mungo Man's DNA suggested his maternal lineage had diverged from the human family tree earlier than those of all people on the planet. Although the results, published in the *Proceedings of the National Academy of Sciences* in 2001, did not imply an Australian Eve, they presented a 'serious challenge' to the out of Africa theory, the team said.

The research, which sent shockwaves through the scientific community and made headlines world wide, deepened ideological divisions on many fronts. The study on the remains of the man roughly contemporaneous with and possibly a close relative of Mungo Lady was of great personal interest to some Aboriginal people curious about their roots. But it came as the campaign for the repatriation of indigenous skeletons from around the world was gathering momentum, a campaign that has aroused controversy over genetics and other research. Continuing controversy over research on ancient remains and modern indigenous people casts a pall over big genetics projects.

And the results went to the heart of the question of the identity of the first Australians, a question at the centre of the debate over human evolution. Were the ancestors of the Aborigines part of a recent African exodus of *Homo sapiens*, or did they evolve from older *Homo erectus* in Indonesia or China? The Mungo Man research intensified the conflict between multiregionalists and Africanists, and the discovery of the Indonesian hobbits, *Homo floresiensis*, would later be a cue for a reprise of the arguments. The 1-metre-tall hominin, with a skull smaller than a chimp's, found in 2003 at Liang Bua on the island of Flores was classified by its discovery team as a member of the genus *Homo*. At just 18,000 years old, the classification upset the multiregionalist scheme because it put two human species in the same place at the same time. Alan Thorne and colleague Maciej Henneberg, of Adelaide University, have been spearheading the attack on Hobbit's classification.

The arguments have been echoed in Europe, where Neanderthals are pivotal to the wider debate.

Multiregionalism was once the dominant paradigm in palaeo-anthropology, but by the time the PNAS paper was published it was a minority view, at least in its extreme form. Multiregionalists argue that *Homo sapiens* evolved simultaneously on various points on the globe from migrants, classified by some Africanists as *Homo erectus* and by others as *Homo heidelbergensis*, who left Africa more than a million years ago. Interbreeding with later migrants nudged our species along the same evolutionary route. Multiregionalists are 'lumpers', arguing that *erectus*, *heidelbergensis* and *sapiens* were everywhere, and always, the same species. To them, regional differences between populations arose at least a million years ago and some have been preserved, accounting for the physical differences between populations today.

The rival out of Africa model gained ascendancy from the late 1980s with strengthening fossil and DNA evidence for a recent African ancestry for all people on the globe. Africanists contend that our species evolved in Africa, probably less than 200,000 years ago, and dispersed across the planet, overlapping with and then replacing one or more other hominin species descended from the earlier migrants. They say that racial differences are only skin deep. In a presentation to science journalists, palaeoanthropologist Peter Brown, renowned for describing *Homo floresiensis*, demonstrated the close similarities between the skull of former Australian politician Pauline Hanson and that of a Chinese woman. Brown, of the University of New England, is one of the leading Africanists in Australia. Another is his former teacher, the quietly spoken ANU don, Colin Groves, an active member of the Australian Skeptics, anti-creationism campaigner, and mentor to most of the palaeoanthropologists not trained by Thorne. Groves, who is described by Jane Goodall as the world's foremost primatologist, has discovered several new primate and non-primate species, living and extinct, some of them co-described with Tim Flannery. He is active in the campaign to save embattled wild populations of chimps, gorillas, orangutans and other primates.

Australia's palaeoanthropologists—a community so small it could hold a conference in a telephone box—are divided evenly between Africanists and multiregionalists.

The Willandra Lakes remains and Victoria's Kow Swamp skeletons, also said by the authors of the PNAS paper to have yielded DNA, have figured prominently in a fiery debate.

The multiregionalists argue that many of the Willandra Lakes people were 'gracile'—slightly built, with the delicate facial features of modern humans, while the Kow Swamp people were 'robust' with thick skull bones, flat receding foreheads and prominent brow ridges. Thorne has argued that the differences testify to two waves of migration into Australia. The Kow Swamp and other robust people, from Cohuna in Victoria, Coobool Creek, southern New South Wales, and Cossack, Western Australia, descended from Indonesian *Homo erectus*. Much of the argument rests on comparisons with the *Homo erectus* skeletons from Indonesia. However, Brown and Groves counter that the robust skeletons are within the normal range for Pleistocene Australians. The receding foreheads, Brown contends, were the result of artificial cranial deformation, a practice common in other populations. And the big jaws and teeth in the Kow Swamp specimens were adaptations to eating unprocessed food, write Groves and David Cameron, formerly of the University of Sydney, in their book, *Bones, Stones and Molecules*.

The other wave of immigrants—'graciles'—were from China, the multiregionalists say. At just under 1.5 metres tall, and lacking a big brow ridge, Mungo Lady is said to epitomise gracile morphology.

Mungo Man, known in the scientific literature as Willandra Lakes Hominid 3, or WLH3, is also said to be gracile, but Brown has disputed this:

> Alan Thorne has argued that the skeleton was of a male, that it was anatomically modern and that it was particularly gracile in build compared with some other Pleistocene Australians from Kow Swamp … This information has been used to support Thorne's belief in the migration of two distinct groups of people to Australia during the Pleistocene. Modern Aborigines were argued to be the descendants of these two groups of Asian immigrants … The information that has been published suggests that LM3 would have been approximately 170 centimetres tall, was robustly built in terms of bone thickness and skeletal dimensions, and due to poor bone preservation is of indefinite sex. In comparison to relatively large-bodied people who lived in south-eastern Australia during the Pleistocene, LM3 was either a tall woman or a shorter than average man.

It is common for physical anthropologists from opposing theoretical camps to come up with different classifications for the same skeleton, no matter how exacting their anatomical measurements. Both the Africanists and the multiregionalists have amassed anatomical data to support their cases, but they argue over the interpretation. The problem is exacerbated by a patchy fossil record with incomplete skeletons, and palaeoanthropologists have drawn a jungle of family trees, or bushes, especially in their literature on Africa. What the multiregionalists see as 'missing links' between *Homo erectus* and *Homo sapiens*, the Africanists see as species at the end of evolutionary lines that went extinct when the moderns entered the scene.

In the late 1980s, while the physical anthropologists argued about sizes, shapes and statistics, the geneticists were just starting to throw DNA into the argument. Their exciting new approach was hi-tech, deploying sequencing equipment and supercomputers that made the callipers of the anatomists look archaic. They drew on the so-called 'hard sciences'—chemistry and mathematics—in work that seemed to contrast markedly with the uncertainty of anatomy. It had poetic appeal, too: our evolutionary past was encoded in the blood pulsing through our veins. It was simply a matter of deciphering the genetic code. The field carried its own uncertainties, however, and was just as open to interpretation.

Genetics had its first big hit in prehistory with the 'Mitochondrial Eve' paper published in *Nature* in 1987, which claimed to trace the genetic heritage of all people on the planet back to a single African mother. The research, by Rebecca Cann, Mark Stoneking and Allan Wilson, of the University of California, Berkeley, focused on variation in the small mitochondrial genome, now known to have just 16,569 base pairs and a mere 37 genes. The team compared sequences, mostly from samples extracted from placentas, of 147 people from five populations, including Australian Aborigines and New Guineans. Cann's group made a sensation when it declared: 'All these mitochondrial DNAs stem from one woman who is postulated to have lived about 200,000 years ago, probably in Africa.'

The research raised the profile of genetics, and subsequent studies using more refined methodology bolstered the Africanist case. The strongest support came from research, led by Peter Underhill, of Stanford University, sequencing the Y chromosome. The work was published in *Nature Genetics* in 2000, just before the Mungo Man DNA paper appeared. Only males have a Y chromosome among their 46, and they inherit it from their fathers, so Y chromosome DNA reveals paternal lineages. The Underhill study showed that the maternal and paternal lineages stretched back to Africa. However, the research on DNA from 1,000 men from 22 regions suggested that Y chromosome Adam post-dated Mitochondrial Eve, by then thought to have lived about 143,000 years ago, by 84,000 years. The researchers argued that several versions of Y chromosome DNA were around at the time of Eve, with one later spreading through the population and coming to dominate 84,000 years later.

In 2001, however, supporters of the rival multiregional model were beating the Africanists on their home turf—the genome. A group of anatomically modern humans like Mungo Man had ventured into Australia carrying mtDNA different from the mother of all mothers.

That study built on postgraduate work of Thorne's student, Greg Adcock. Respected geneticist Simon Easteal, a former Africanist who had made a late conversion to multiregionalism, was also in the group, as were scientists from, curiously, the CSIRO Plant Industry Division.

The team claimed to have recovered mtDNA from the bones of WLH3 and nine other ancient Australians. Three of them were also from the Willandra Lakes. One had been AMS radiocarbon dated by Gillespie and Fifield to 5,000 years, and another to less than 100 years old. The third had been estimated to be a few hundred years old by palaeoanthropologist Steve Webb. The other six were from Kow Swamp, and had radiocarbon dates of 10,000 to 15,000 years, although OSL dating by Melbourne University's Tim Stone and Matt Cupper suggests the ages could be 19,000 to 26,000 years.

The team also compared Mungo Man's mtDNA with that of 45 Aborigines living today and 3,400 people from populations around the globe, and with sequences from two Neanderthals, a pygmy chimp and a common chimp. They found that Mungo Man's lineage split off from the family tree earlier than those of all other modern humans sampled, and he had a sequence not found in the mtDNA of any other sample. The sequence does occur in many people around the world, but today it is found only on chromosome 11 of the nuclear genome. At some time in the distant past, this 'insert sequence' had migrated from the mitochondria and was now part of a mixed maternal/paternal inheritance, the scientists said. Its presence in Mungo Man's mtDNA was seen as a palaeoanthropological smoking gun. The scientists argued that Mungo Man's people carried two types of mitochondrial DNA—the ancient kind and the contemporary kind. The latter spread by Darwinian natural selection throughout the population, so it must have conferred some advantage, perhaps to do with powering the brain.

'Our data present a serious challenge to interpretation of contemporary human mtDNA variation as supporting the recent Out of Africa model,' the paper said. Easteal was cautious in interviews during the ensuing media frenzy. 'Our results don't mean that modern people emerged from Australia any more than deep lineages in Africa suggest a recent origin of people on that continent,' he said. 'We cannot make judgements based on one gene. We can therefore challenge but neither confirm nor refute the out of Africa model on the basis of our data.'[1] Thorne, a former journalist who studies snakes as a hobby, gave the results more of a multiregionalist interpretation. And he was backed up by palaeoanthropologist Milford Wolpoff, a well-known multiregionalist colleague at the University of Michigan. Wolpoff was quoted by Reuters as saying: 'There never was a marauding band of Africans. It certainly means that the Eve theory, the replacement theory, seems to be wrong.'

Peter Brown said ancient DNA was unlikely to have been preserved in the hot, dry conditions at Lake Mungo. And the ANU's Colin Groves said discovery of an ancient, extinct mtDNA sequence would not weaken the out of Africa model anyway. Many types of mtDNA would have existed in our species' distant past. Most went extinct when women carrying them had only sons.

Other scientists criticised the results on the grounds of the phylogenetics, the complex field involving finding the most parsimonious family tree to fit the data. The team had used a supercomputer to construct its phylogeny, but these days the task can be done on a desktop or even a laptop. Writing in the journal *Archaeology in Oceania*, John Trueman, a geneticist at the ANU, said Mungo Man sat alongside all of us on one branch of the human family tree. Don Colgan, a geneticist at the Australian Museum, said different analytical approaches delivered different results, some with Mungo Man on our branch of the tree.

And in a letter to *Science*, six researchers, including Alan Cooper, then at Oxford University, and palaeoanthropologist Chris Stringer, of London's Natural History Museum, said they had constructed different family trees, one putting Mungo Man well within the modern human range simply by including more Aboriginal and African sequences in the analysis. The research team defended its results in a reply in the same journal. And Adcock told the *Australian Financial Review* that the sequence was likely to be a regional variant. If it were part of an African exodus, he said, it would probably have turned up somewhere in the world by now.[2]

Cooper and colleagues also charged that the research team had failed to follow standard procedures in handling ancient DNA—cloning and replication by an independent laboratory—aimed at demonstrating that the sequences were authentic, and not the result of contamination by modern DNA or of DNA damage. Contamination has caused embarrassment in the past, especially when American scientists claimed they had recovered DNA—as in *Jurassic Park*—from dinosaurs, only to find it was human. Other suspect results include Egyptian mummies, and termites and weevils found in amber.

The contamination problem is worsened by the polymerase chain reaction, which has been a boon to ancient DNA research, amplifying minute DNA samples, but a double-edged sword. Scientists heat the sample to separate the DNA strands, and add primers—short nucleotide segments—along with the ACGT bases and a form of the enzyme polymerase extracted from a thermophile—a bacterium that can take the heat. When the mixture cools, the primers bond to the ends of the DNA fragment to be amplified, with one primer latching onto each strand. The polymerase adds the bases to the primers, using the sample's single strands as templates and following the base pairing rules. The process is repeated, with every cycle doubling the number of DNA fragments. Thirty PCR cycles pump out more than a billion fragments.

A gram of old bone typically contains 10,000 to a million copies of the DNA sequence of interest. After a successful PCR amplification, a single copy of DNA can end up as 10 billion copies of the target sequence in only 10 to 15 microlitres, the volume of a small raindrop. However, DNA contamination from people who have handled the specimen is also amplified. Although the geneticists working in the laboratory sequence themselves to ensure their own DNA doesn't wind up being the subject of their papers, it is often impossible to sequence everyone who has handled the material, especially if standards in the field are lax. The DNA of archaeologists and excavation teams often appears in the analysis, especially if the sample is washed, a practice that ferries contaminants deep into the material. Modern DNA outstrips the ancient DNA and often it is impossible to detect contamination via the sequence results. 'Archaeologists like to handle things, and I think they get very excited when they find things—maybe they drool a bit more,' says Cooper. 'As a result we can find an enormous amount of

DNA from the people working on that material. We've also found a lot of Chris Stringer. He was either on the lineage that gave forth to all of humankind or he's touched a lot of bone.'

'You're all very dirty,' Cooper told archaeologists and palaeontologists in a lecture aimed at lifting standards of sample collection for DNA analysis. 'There's a pool of your own DNA around you now. If you can't rule out contamination as the source of the results, you undermine any faith you have in the analysis,' he warned.

When a contaminated ancient DNA sample is sequenced directly after amplification, the result is a melting pot sequence of all the individuals represented in the sample. Cloning overcomes the problem. It involves lobbing the PCR products into bacteria genetically engineered to take up only one DNA molecule each. The bacteria replicate the DNA sequence as if it were part of their own genomes. Single bacteria are cultured and clones from each colony amplified and sequenced. Cloning also reveals post-mortem DNA damage.

The Mungo Man research team isolated less than 2.3 millionths of a gram of DNA from each gram of bone sample. The researchers reported working under a regime of ritualistic cleanliness to avoid contaminating the samples with their own DNA. Only two scientists handled the remains during the experimental work—Greg Adcock and Alan Thorne—and they sequenced themselves to ensure that their own genetic fingerprints did not figure in the analysis.

'None of the samples had been handled by either Aboriginal or non-Aboriginal people before extractions began,' the team said in its response to critics in *Science*. However, the list of researchers who handled the skeleton between its discovery in 1974, when no-one had conceived of extracting DNA from such old bones, and Adcock's work in the 1990s, reads like a Who's Who of Australian prehistory. It was a genetically diverse population, ranging from geologists through archaeologists to palaeoanthropologists, and from multiregionalists to latent Africanists.

Cooper and colleagues also argue for replication at an independent laboratory. Samples were sent to an independent laboratory overseas for analysis in 2005, but at the time of writing the research had not been published in a major journal.

Despite the sensitivities of DNA research, some has been conducted on Australian samples, and the arguments over the palaeoanthropology and dating continue.

If *Homo sapiens* evolved in Africa perhaps up to 200,000 years ago, and dispersed between 50,000 and 80,000 years ago, the last big question is what took our species so long to get out, says geneticist Peter Forster of the UK's Anglia Ruskin University. He describes the multiregionalism/out of Africa argument as 'yesterday's debate'. Another question is the route the pioneers took—perhaps via the southern Arabian peninsula to South-east Asia and Australia, or perhaps northwards via the Middle East.

What did they find when they got to South-east Asia—*Homo erectus* or a tropical paradise that was *terra nullius*? The question has implications for the date of colonisation of Australia and the ancestry of the Aborigines. The deep background could lie in another adventure—dubbed Out of Africa I—by primitive hominins almost 2 million years ago. Asia might have played a bigger role in our species' remote evolutionary past than previously thought, and when the ancestors of the Australian Aborigines reached South-east Asia, they might have been coming home.

16 Roots

Is this really the cradle of humankind?

'They call it that but I don't like the term. If they want to talk about cradle of humankind, they have to include the whole of Africa, and even that might not be correct. We might find in Asia, in the near future, older hominid fossils than in Africa.'[1]

Ron Clarke, a professor at the University of Witwatersrand, was descending the stairs inside the giant cave of Sterkfontein in South Africa's Gauteng Province. The route to the site, 50 kilometres north-west of Johannesburg, traverses a landscape of nondescript hills rising from open pastures with scattered shrubs and trees. Sterkfontein is nestled in a 50,000-hectare valley known as the Cradle of Humankind. Along with Swartkrans, Kromdraai and Makapansgat, it was put on the UNESCO World Heritage list in 1999. Clarke and other renowned anthropologists, including Phillip Tobias (also from Witwatersrand), Robert Broom, Raymond Dart and Alun Hughes, devoted their lives to studying Plio-Pleistocene hominin fossils reaching back to the time when the first bipedal 'ape-men' appeared. More than 2 million years of hominin activity is recorded in these caves, home to 40 per cent of the world's human ancestor fossils.

The steps of Sterkfontein penetrate deep geologic time to about 2.6 billion years ago, when dolomite rock formed on the floor of an ancient sea that covered the interior of South Africa. Hundreds of millions of years later, after the ocean had retreated, ground water penetrated the subterranean formation, sculpting the caves as it dissolved the dolomite. Further dissolution of the rock along vertical cracks, and land erosion, eventually opened the caves up to the light. Animal and hominin remains, along with sediments, were washed in through steep shafts, and were cemented by calcium carbonate into a solid breccia rock matrix.

The Sterkfontein formation lacks the uniform stratigraphy of sedimentary rocks, so to give it time order the system was divided into components, from Member 1, the oldest, to Member 6, the youngest. The deposits were first dated indirectly through the appearance and disappearance of fauna from the fossil record and the evolutionary changes in the animals' anatomy. Scientists compared the fossils of animals with other East African ones dated directly. Direct and absolute dating of Sterkfontein has been difficult, with the oldest remains out of the reach of radiocarbon and luminescence techniques. And there is no volcanic material lending itself to potassium–argon dating, a method based on the radioactive decay of potassium-40 to argon-40, which was used to date Lucy the australopithecine and other renowned hominins in eastern Africa. Recent technological advances have opened the way to palaeomagnetic dating. Geologist Tim Partridge, of Witwatersrand University, has been studying Sterkfontein for 30 years and is still grappling with its stratigraphy and chronology. 'The mark of the Gilbert/Gauss polarity reversal at 3.58 million years ago is evident at the bottom of the sequence for Member 2, while the Reunion Event at 2.14 to 2.15 million years ago gives the chronology of the younger Member 4,' said Partridge, referring to geomagnetic excursions used widely by dating experts.[2]

More than 500 hominin skull, jaw, tooth and post-cranial fossils have been recovered from Sterkfontein's deposits spanning the past 3 million years, a critical period for the metamorphosis from 'ape-man' to *Homo*. Thousands of stone tools recovered include the earliest handiwork of our genus, dating to about 2 million years ago. The most recent deposits contain tools from the time when, according to the Africanists, *Homo sapiens* was emerging in Africa, between 200,000 and 100,000 years ago. Fossil remains of plants and animals paint a picture of the landscape.

'Here is where Dr Broom and John Robinson found Mrs Ples,

the nickname for *Plesianthropus*, in 1947,' says Clarke, pointing to a section of Member 4, site of the discovery of a 2.15-million-year-old fossil australopithecine, one of our most distant hominin relatives. 'During those years, 1936 to 1947, miners were working in this region using dynamite and found several fossils of australopithecines and animals.' Australian expatriate anatomist Raymond Dart, of the University of Witwatersrand, in

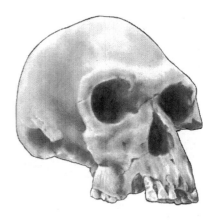

H. habilis

1925 first recognised and described an australopithecine skull from Buxton Limeworks in the North-West Province of South Africa. He named the 'Taung child' *Australopithecus africanus*, or southern ape of Africa. Despite its small ape-like brain, Dart's hominin, the first of many australopithecine species to be discovered, was walking upright and had other human features.

Walking through Silberberg Grotto, Clarke points out the bones of Little Foot, another *Australopithecus*. Tim Partridge got a palaeomagnetic age of 3.2 to 3.6 million years, and a cosmogenic age of 4 million years, making Little Foot the oldest *Australopithecus* in Africa, older than Ethiopia's Lucy, the most famous member of the genus. Other scientists, including Clarke, dispute the date.

Australopithecines weighed 25 to 50 kilograms and stood up to a little more than 1 metre tall. Their cranial capacity was just 400 to 500 cubic centimetres, similar to chimpanzees' and hobbits'. They were not much smarter than chimps, some populations of which use stone hammers and anvils to crack nuts. Member 5, dated from 2 to 1.7 million years old, contains Oldowan stone tools, the mark of our genus, *Homo*. It records a drier period, and it is here that a skull considered by Tobias to be *Homo habilis* was found in a breccia, with the rock's calcified cave sediments also holding 3,000 Oldowan artefacts. *Homo habilis*, or Handy Man, is thought to be the first member of *Homo*, but some scholars lump it in with *Australopithecus*.

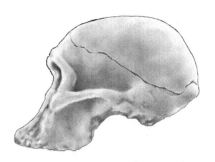

A. *africanus*

'This was a challenging moment for hominins in Africa who needed to change diet and way of life in order to survive,' Tobias later said of the find. 'An early form of *Homo* with larger brain and stone tools suddenly appears in the Sterkfontein record in response to environmental change,' says the eminent scientist, who describes himself as a neo-catastrophist, emphasising the role of environmental disasters in the unfolding of human evolution.[3] Clarke classifies the specimen as a male *Australopithecus africanus*, however.

H. habilis was about the same size as *Australopithecus* but was fully bipedal, with shorter toes, not adapted to climbing trees, and a larger brain with structures on the left hemisphere that Tobias associated with language. Not being big on nuts, *Homo* had smaller jaw muscles than australopithecines, and this might have cleared the way to thinner skulls and bigger brains. It is more likely, however, that the force driving the evolution of the genus that would one day sequence its own genome was climate change, which transformed parts of Africa from tropical forests to savannahs.

'I don't think *Homo* necessarily evolved from any of the *Australopithecus* species we know,' says Clarke, who has been in charge of Sterkfontein since 1981. 'It could have evolved from a similar species, but not necessarily from one we have found.' He sees australopithecines everywhere, and believes they were all over Africa and probably in Asia. He believes australopithecines have also been found in China, but they have not been classified or described in the literature. Some scientists argue that australopithecines and *Homo habilis* beat *Homo erectus* to Asia. Some contend that the human genus could have evolved in Asia, where sites have been found with stone tools older than the accepted date for out of Africa I. Perhaps, australopithecines and/or *habilis*-like hominins were living in the vast savannahs stretching from western Africa to China 3.5 to 2.5 million years ago, but their remains have not been preserved or found (at least officially).

About 1.5 million years ago, more sophisticated 'Acheulean' stone tools appeared in the Sterkfontein breccia. The technology, which spun off from Oldowan tools and is epitomised by the handaxe, has been attributed to our close ancestor *Homo ergaster*. It was based on raw materials sourced within 20 kilometres of the hominins' home ranges. It expanded rapidly throughout the Old World but advanced technologically only slowly between 1.5 million and 300,000 years ago.

'By about 1.5 million years ago,' continues Clarke, 'hominins were depending on the use of stone tools as a regular part of their strategy for survival.'

The details of *H. ergaster*'s ancestry remain uncertain. The species appeared suddenly in Africa about 2 million years ago, and includes the Kenyan Nariokotome boy. Specimens found in the 1960s and 1970s were classified as *H. erectus*, and some scholars still lump them in with that species. However, Colin Groves and Czech

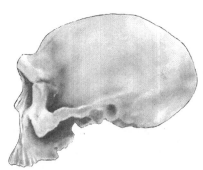

H. ergaster

biologist Vratislav Mazák in 1975 assigned the fossils to a new species, *H. ergaster*. At 1.75 metres tall, males of this species would have been almost twice the weight of an australopithecine and had a brain capacity of 900 cubic centimetres, and modern body proportions differing by only 20 per cent between the sexes. A recent discovery in Kenya modified ideas about how *ergaster* may have evolved from *habilis*. A 1.44-million-year-old *H. habilis* jawbone and a 1.55-million-year-old '*H. erectus*' skull from the Koobi Fora formation, east of Lake Turkana, put the two species side by side for more than 500,000 years. Clearly, *H. ergaster* did not evolve from *H. habilis* 'en bloc', but from only a small, presumably isolated, population of it. The adult skull is small, and its discoverers argue that *ergaster/erectus* had a bigger size range than previously assumed. The discoverers also identified 'Asian' characteristics, suggesting greater affinity between the African *ergaster* and its *erectus* counterpart in Indonesia, home of the *erectus* type specimen.

Perhaps the more advanced *Homo* species evolved in Asia and dispersed back to Africa. The Natural History Museum's Chris Stringer cautions against dismissing the scenario.

In the Lincoln Cave, north of Sterkfontein's Member 5, deposits from the Middle Stone Age, 300,000 to 50,000 years ago, carry the first signs of modern humans. During this period, large cutting tools were replaced by smaller tools based on Levallois core technology. This technology, involving the production of flakes with precise shapes from pre-prepared stone cores, is named after Levallois-Perret, a suburb of Paris where such tools were first found. Composite tools that had already appeared during the Acheulean technological phase became increasingly complex, requiring design, standardisation and the import of raw materials, sometimes from quarries hundreds of kilometres away. Complex language would have been critical to the manufacture of these tools, which diversified into regional styles.

The fossil record confirms that by 100,000 years ago anatomically modern humans were in Africa and the Levant, leaving their traces at sites, including Border Cave and Klasies River in South Africa, Omo-Kibish in Ethiopia, and the Qafzeh and Skhul Caves in the Levant. When they became behaviourally modern, with the capacity for complex language, planning and using symbols, is more difficult to pin down. They probably had the neural 'hardware' and perhaps even 'software' for complex language by 100,000 years ago, but they might not have used this capability until tens of millennia later. Perhaps a genetic change amid a population expansion propelled the transition to modern behaviour.

Stone tools underwent further development, and the moderns used bone, ivory, shell and antler to make harpoons, needles, buttons and ornaments. They made fibre nets and bags, wore animal skin fashions, harvested seafood, ground seeds, painted their bodies and drew with ochre on cave walls. They had the baggage for a great trip that would take them to Asia, Australia, Europe and finally to America and the Pacific. A new age of symbolism and self-awareness was dawning. Modern humans left Africa with their weapons, words and wiles.

***

'I'm in no way suggesting a controversial age for modern humans in Java based on the evidence of one tooth,' timelady Kira Westaway told the 2007 International Union of Quaternary Research meeting in Cairns. She was referring to a precious *Homo sapiens* pre-molar recently rediscovered in a museum by Paul Storm, of the National Museum of Natural History in the Netherlands. It was found in 1939 in the yellow breccia of Java's Punung 'faunal stage', the second most recent sequence in a series of seven fossil-bearing deposits extending across the island.

The faunal stages record the appearance and extinction of animals over the past 1.5 million years, and the Punung deposit paints a picture of wet rainforest when orangutans reigned. It overlies and is therefore younger than the Ngandong faunal stage, which holds the fossils of extinct animals, including *Homo erectus*. Until recently, there were no reliable dates on the Punung stage. Westaway, of the University of Wollongong, has now dated it through the OSL and uranium-series methods to 120,000 to 130,000 years old. At the time of writing, the oldest modern human skeleton found in South-east Asia was a 45,000-year-old skull from Borneo's Niah Cave. Westaway's team claims its work, published in the *Journal of Human Evolution*, challenges young Ngandong dates, pushing back the date of *erectus* extinction in South-east Asia, and perhaps the arrival of modern humans there.

Westaway's study formed part of wider research by Richard Roberts, Mike Morwood and colleagues retracing the steps of the first Australians as they passed through Sunda (the land including mainland South-east Asia and western Indonesian islands during the Pleistocene) on their way from Africa to Sahul (the combined Australia–New Guinea–Tasmania land mass). The team was attempting to sort out the confused chronology of Indonesia, a region so critical to the arguments about human evolution. The big questions hinge on the timing of the extinction of *Homo erectus* and the arrival of *Homo sapiens*. While the palaeoanthropologists focused on the comparison between Indonesian *Homo erectus* and robust and gracile Pleistocene Australians, Morwood's team was taking a different tack—seeing if there was a replacement of our more primitive cousins.

Climate change periodically opened corridors between East

Africa's Great Rift Valley and Asia. Primitive hominins, such as *Homo ergaster*, or *Homo erectus*, which evolved in Africa about 2 million years ago, dispersed into Eurasia. Even more primitive hominins, such as australopithecines, might also have left. *Erectus*'s dispersal was swift, with remains from Dmanisi in Georgia considered *erectus*-like hominins by some and dated to 1.8 million years old. Scientists have obtained similar dates for Javanese *H. erectus* from sites including Sangiran. There, fossils like Sangiran 17 conforming to the popular *erectus* image of a hulk with a thick skull, flat forehead and projecting face, have been dated to about 1 million years old or slightly less. The Modjokerto child skull from Java dates from 1.8 million years ago. Meanwhile, stone tools from Majuangu, northern China, have been dated to 1.66 million years. Some scholars insist that *erectus* was a seafarer, perhaps even sailing to Australia, but many consider the notion a bit 'out there'. Steve Webb writes that 'these people [*erectus* living in South-east Asia] or their descendants were the most obvious to be the first Australians, long before the "advent" of anything called anatomically modern human'. Chris Stringer says pre-modern humans in Europe, at least, were incapable of making sea crossings on boats. 'There's no clear evidence that people crossed the Strait of Gibraltar, for example, by any kind of watercraft before modern people did, and the Mediterranean islands ... as far as we know never had people on them at the time before modern humans.'

Scientists dated the first appearance of *erectus* in South-east Asia by deploying the potassium–argon method on pumice assumed to be associated with the hominin remains. Volcanic glasses in pumice have a lot of potassium, and the known rate of decay of the radioactive isotope potassium-40 to argon-40—a process with a half-life of 1.28 billion years—can be used to date the rock. The daughter product is a gas and does not exist in the pumice or other volcanic ejecta when the rock forms from lava but builds up as the parent isotope decays. Scientists work out the concentration of argon-40 by heating the sample and using a mass spectrometer to count the atoms released. Dating accuracy has recently been improved by extracting the argon atoms from single crystals picked from the pumice, using a laser. The method works best on samples at least 100,000 years old, but it can be used on samples as young as 10,000 years as long as they are rich in potassium. The University of California, Berkeley, laboratories used the method to date the Vesuvius eruption of 79 AD. The method was applied in the 1970s to date the first australopithecine finds—specimens more than 3 million years old—and can be extended to chronologies close to the age of the Earth—4.3 billion years.

Controversial dates obtained in 1996 by Carl Swisher, of Rutgers University, suggesting *H. erectus* held on in Indonesia until very recently have fanned the fire of evolutionary theory. He got dates of 27,000 to 53,000 years for an *erectus* fossil site on a terrace above the Solo River, near the central Javan village of Ngandong. The remains were discovered in 1931, four decades after the Dutch surgeon Eugene Dubois, on the trail of the erroneously hypothesised 'missing link' between apes and humans, found the *H. erectus* type specimen at nearby Trinil. Dubois christened 'Java Man' (Trinil 2) *Pithecanthropus erectus*. But in the 1950s the more slightly built Trinil specimen was lumped with 'Peking Man' into a new single species, *Homo erectus*.

Twelve hominin skull caps and other bones were found at Ngandong in the following years during excavations organised by the Dutch Geological Survey. The archaeologists did not find teeth, the most reliable material for direct dating, and the keeper of the fossils, the late Teuku Jacob, of Gadjah Madah University, forbade the extraction of skull samples. Swisher applied ESR and uranium-series dating to old bovid teeth from the hominid-bearing levels. His ages are contested because they are up to 400,000 years younger than previous dates for these hominins, suggesting *erectus* lasted on Java 250,000 years longer than on the Asian mainland and one million years longer than *H. ergaster* in Africa. Some scientists challenge the dates on technical grounds. Still others question the tooth samples' association with the remains. Roberts, Morwood and Westaway say: '... the present lack of a robust chronology for the major turning points in the South-east Asian cultural and evolutionary sequences makes it futile to speculate on whether *Homo erectus*, *Homo sapiens* and *Homo floresiensis* ever came into contact.'

Colin Groves says Westaway's team's Punung fauna results could be documenting the movement of *erectus* in sync with climate change rather than the extinction of the species. The hominins might have retreated to the drier forests and open country favoured by humans when rainforests claimed much of the island at the end of the second-last ice age. If Westaway's results do document the exit of *Homo erectus* by 120,000 years ago, they will eliminate the hominins as the progenitors of the robust Australians. Thorne has said previously that if Swisher's younger dates stand, the population represented by the Ngandong specimens 'must have ancestors who would still be excellent candidates for the migrations from South-east Asia that formed the basis of the earliest Australian population'.

A team led by Yuji Yokoyama, of France's Natural History Museum, muddied the waters in 2008 when it published uranium-series dates of 40,000 to 70,000 years on the Ngandong fossils.

Genetics research is running in counterpoint to the palaeo-anthropological and dating work as scientists put together the itinerary of the modern human travellers. One of the most closely studied populations is the Andaman Islanders. The land of 'head hunters' is how Marco Polo described the Andaman Islands in the thirteenth century. The 'Negritos' on the islands in the Bay of Bengal have for centuries had a reputation for killing anyone landing on their territory. The island-ers' ferocity towards outsiders protected them, but their future is now uncertain.

In 1858, the British managed to gain a foothold, establishing a penal colony. The colonial government attempted to 'civilise' and impose change on the Andamans, triggering a population crash. The largest group, the Onge, has diminished greatly from 1,000 in 1901 to fewer than 100 now. The islanders differ markedly from other Asians, looking more like pygmy groups in Africa. There are similar groups in the Philippines, Malaysia, and perhaps in the jungles of Andhra Pradesh, in India. Who are they? Are they relict popula-tions connected to the early dispersal of *H. sapiens* from Africa to Australasia?

More than 40 years ago, Italian geneticist Luigi Luca Cavalli-Sforza of Stanford University pioneered the reconstruction of human dispersal using modern genetics. His method was based on 20 genes associated with five blood groups from 15 indigenous populations.

The first evolutionary tree, which took shape when a then state-of-the-art computer at the University of Pavia slowly crunched all the data, showed that the maximum genetic distance was between African groups and Australian Aborigines. These early data could not pinpoint the root of the tree, and Cavalli-Sforza originally suspected an Asian origin. While an African origin is now generally accepted, the date and route of human dispersal remain controversial.

Early out of Africa models assumed a dispersal from sub-Saharan Africa northwards via the Levant and the eastern Mediterranean coast about 45,000 years ago. In later models, an earlier dispersal around 60,000 to 70,000 years ago was suggested, with the migrants coasting the Arabian peninsula and India to South-east Asia and Oceania. Mitochondrial DNA analyses found evidence for a single dispersal from Africa, but these studies still could not trace the route.

A vast desert during the last Ice Age blocked the northern route through the Sinai. That raises the question of how modern humans got to the Levant 100,000 to 130,000 years ago. Bones and teeth have been dated to that age through the uranium-series and ESR methods. Palaeoclimatological records in speleothems from the Negev desert suggest a fairer climate in the Sahara between 140,000 and 110,000 years ago might have cleared the way for an episode of human dispersal then. The travellers probably never ventured beyond the Levant, however. They disappeared from the fossil record about 80,000 years ago after a brief overlap with the Neanderthals, and were probably forced to lower latitudes as northern Eurasia got colder.

More recent mtDNA studies support the hypothesis of a single exodus from Africa, probably via the Bab-el-Mandeb Strait in the southern part of the Red Sea, 60,000–70,000 years ago.

The analysis of complete mitochondrial DNA sequences from relict populations in Eurasia suggested they arrived in India 66,000 years ago, and a study of stone tools from a site in Andhra Pradesh in India, dated by a team led by Cambridge University archaeologist Michael Petraglia, supports the genetics. The site, dated with OSL, contains artefacts made as early as 77,000 years ago, and is claimed to be the oldest *Homo sapiens* site in India. The stone tools were found in sediment layers above as well as below fine ash from the super-eruption of the Toba volcano in Sumatra, the largest volcanic explosion in the past 2 million years, which hurled nearly 3,000 cubic kilometres of rock and ash into the atmosphere. Still, the modern humans survived and continued their trek to Australia. Now the Toba ash is a big money-spinner for the locals, who sell the fine powder, widely used as a metalware polish, in the markets.

***

The DNA of South-east Asians suggests that their African ancestors comprised a group of 700 individuals, according to a recent genetic study involving 52 worldwide populations. They may have been the descendants of the people who left Middle Stone Age handaxes and obsidian flakes in sediments in Abdur, Eritrea, and sites in South Africa during the last interglacial (OIS 5).

After the last interglacial, about 120,000 years ago, the planet started on a cooling trend that lasted 100,000 years. Sediment records from Lake Malawi point to severe aridity in tropical Africa between 135,000 and 75,000 years ago. But the climate improved about 70,000 years ago, possibly boosting the population size and launching the people on a hunt for new resources. Generation after generation, *Homo sapiens* edged along the tropical coast of the Indian Ocean. They had developed behaviour that allowed them to out-compete other *Homo* species already in the new lands. Around 50,000 years ago, warm interstadial conditions opened the way for the dispersal of these behaviourally modern humans from western Asia and low latitudes into Europe and north Asia. Modern stone tools and carved ivory ornaments were found at a site on the Don River in southern Russia in sediment layers dated with OSL to between 45,000 and 42,000 years ago.

In one of the biggest analyses of modern DNA so far, an international team including Georgi Hudjashov and Peter Forster compared 172 Aboriginal and New Guinean mtDNA samples, and 522 Y chromosome samples, with populations from around the world.

The team analysed several parts of the mitochondrial genome, including stretches called hypervariable segment 1 and hypervariable segment 2 in the control, or non-coding, region. These fast-mutating regions, which don't carry genes, are often targeted in population genetics. The scientists also analysed several markers—known mutations—that position people on the global human family tree. The number of genes and DNA segments considered made the study one of the most highly resolved ever. All of the DNA, mitochondrial and Y chromosome, told the same story.

The results, published in the *Proceedings of the National Academy of Sciences*, suggested an African origin for Australian Aborigines, with no contribution to their gene pool by local (South-east Asian) *Homo erectus*. Deep mtDNA and Y chromosome sequences branching between Australia and most other populations around the Indian Ocean suggested a long period of isolation after initial colonisation, the scientists said.

Australians were found to be most closely related to the New Guineans, and both probably derived from a single founding population, which according to the team's molecular clock calculations colonised the region about 50,000 years ago. And based on a mutation rate of the complete mtDNA genome of one mutation per 5,000 years, the founders made good time in their journey from Africa via the Indian Ocean coastline. It would have taken only about 5,000 years to make the trip, the paper said.

If they didn't encounter Indonesian *Homo erectus*, they might have run into even stranger hominins. The discovery of the Indonesian hobbits sparked a battle that made the disagreement over the Mungo Man DNA study look tame.

17 Hobbit

After getting a crash course from an Adelaide dentist, Alan Cooper and Jeremy Austin are preparing to drill through the enamel of one of Hobbit's molars to collect a tiny sample of dentine for ancient DNA analysis. The two have a dentist's drill and a portable 'clean box' for the job.

They are in downtown Jakarta at the National Centre for Archaeology (ARKENAS), in a laboratory stacked with hundreds of bags of stone artefacts and the bones of wild pigs, dwarfed elephants and giant rats. The material spans more than 100,000 years of evolutionary history in the strange realm east of the Wallace Line. It has been recovered from Flores over years of excavations by archaeologist Mike Morwood, who coordinates the Australian and Indonesian team that

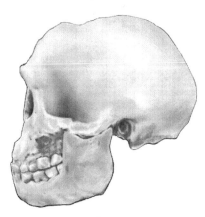

H. floresiensis

discovered the Hobbit specimen, assigned formally by the University of New England's Peter Brown in 2004 to a new human species, *Homo floresiensis*. There has been a lot of grief since then, and the atmosphere is tense after months of uncertainty. Doing research on the Flores material calls for the political skills of a mediaeval pope.

On this day in January 2006, Morwood is in the lab with Wahyu

Saptomo, one of the few team members left at the Liang Bua cave site when Hobbit was discovered in the final stages of a two-month excavation season in 2003. One problem facing the ancient DNA specialists is how to get the sample out without harming the precious type specimen, Liang Bua 1 (LB1). Sample collection is by now a sensitive issue at ARKENAS, after Hobbit was damaged while in the hands of rival scientists according to the discovery team. The other problem is getting a sample free of contamination from modern DNA. Hobbit's teeth are more likely than her bones to yield DNA without contamination from the multitude of people who have handled her.

Another ancient DNA team, led by Svante Pääbo, had collected a tooth sample just a few weeks before. Cooper and Pääbo would race to see if the molecules could settle a debate centred on the bones— whether Peter Brown was right in classifying Hobbit as a new human species, or if instead, as Brown's rivals argue, the specimen was of a modern human affected by microcephaly or some other genetic disease. The question has big implications for the debate over the out of Africa and multiregionalist theories, a debate in which dentistry is a recurring theme.

Since the 2004 publication of the research in *Nature*, the Africanist Brown had been in a bitter dispute with Australian and Indonesian multiregionalists. An academic argument over human evolution had escalated into an international scandal. Lowlights included the snatching of the type specimen, a one-year ban on research, a bitter row in the media between individual scientists and between research institutions, and the exploitation of Australian–Indonesian tensions amid charges of Western scientific hegemony and 'neo-colonialism'. Fanning the flames was the personal politics of the hierarchical Indonesian academic world.

So far, neither team has found any hobbit DNA.

Morwood had been working on Flores for several years before turning his attention to Liang Bua in 2001 as part of his research to find out when modern humans arrived in Indonesia on their way to Australia. But there were even stranger secrets buried in the cave near Ruteng.

Flores, a rugged island 360 kilometres long and up to 70 kilometres wide, lies between Java and Timor. It has attracted all kinds of travellers over the millennia. In more recent times, its mountain lakes, luxuriant rainforests and dramatic volcanoes have drawn only the most ardent adventurers and scientists. Portuguese colonials formed the last big wave of visitors, arriving in 1515. Their missionaries descended on Flores in the 1570s, and the army erected a fort in the centre of the island in the 1630s. The missionaries and the Raja smoothed relations between the Portuguese and the locals, who have subsumed Catholicism into an animistic world view that survives today. Chinese and other Asian populations traded with Flores from the twelfth century. They were attracted to Australia and South Asia also by the trepang (sea cucumber) that was collected in the shallow waters of the region and highly valued on Chinese markets.

In earlier times, during the age of the hobbits, dwarfed *Stegodon* was central to island cuisine, and Morwood's team has found evidence of butchering. Giant rats and Komodo dragons were probably also popular protein sources, with the latter likely to have returned the favour. The giant monitor lizards today provide the unique tourism experience of being prey, attacking visitors in scenes reminiscent of the velociraptor attack in *Jurassic Park*. Rangers on the nearby island of Rinca regaled Tuniz with a story, recounted dispassionately and in minute detail, of a Swiss tourist who succumbed first to blood poisoning and was then eaten by a Komodo dragon.

Volcanoes shaped the landscape and perhaps drove evolution on the island. From the top of Mount Kelimutu is a view of three volcanic lakes of different colours: green, brown and black. An explosive volcanic eruption around 13,000 years ago might have wiped the hominins out around Liang Bua, which has a layer of white volcanic tuff in its strata.

The road to Ruteng from the fishing village of Labuan Bajo on the west coast of Flores is indulgently labelled the 'Trans-Flores Highway' in tourism brochures. From Ruteng, the base of the Australian and Indonesian archaeologists, it's about a one-hour drive along a treacherous road through hills, rice paddies and small traditional villages to Liang Bua, just 14 kilometres away. The journey is not for the faint-hearted. The limestone cavern, possibly the setting for dramatic encounters between different hominin species, penetrates 40 metres into the hillside and is 30 metres wide and 25 metres high at the entrance, with massive stalactites. It was in Sector VII, on the left wall, about 10 metres into the well-lit cave, where *Homo sapiens* became acquainted with their diminutive cousins in 2003. That's where LB1's skull was discovered, when Indonesian team member Thomas Sutikna was supervising local Manggarai villagers as they dug into brown clay with trowels. When their 2-by-2-metre trench hit the 6-metre mark, they saw an object they immediately recognised as a skull. Work slowed as the team carefully recovered the remains—a skull, complete jawbone, teeth, leg bones and pelvis—with sharpened bamboo slivers. It took three days to lift the remains and allow the bones, which had the consistency of wet blotting paper, to harden. The team transported the remains back to ARKENAS for analysis. Then Indonesian lab politics came into play.

Morwood, in Jakarta at the time of the discovery, suspected that the father of Indonesian archaeology, former director of ARKENAS, Radien Pandji Soejono, would hand the remains over to his friend Teuku Jacob of Gadjah Madah University, a veteran palaeoanthropologist and multiregionalist. Jacob, who died in 2007, was not on the discovery team. Soejono was still influential at ARKENAS, from which he had retired in the 1980s. Fearing that the amazing discovery would be lost to science, Morwood flew to Australia and enlisted Peter Brown. They were back in Jakarta in time for the return of the excavation team and its precious cargo. Morwood used the intellectual property agreement between the University of New England and ARKENAS to block the handover.

The remains were unlike anything Brown had ever seen. This tiny hominin walked upright, had prominent brow ridges, a low receding forehead and no chin. Morwood recalls the moment when Brown poured mustard seed into Hobbit's skull to measure her brain capacity. 'There was a stunned silence when this showed that the hominin had a cranial capacity of less than 400 cc,' he says. The figure has since been revised to between 380 and 420 cubic centimetres. For comparison, the mustard seed measure for chimps is 500 cc and 1,400 cc for modern humans. Given that the genus *Homo* had a cut-off point of

500 cc, was this a new genus? The archaeology suggested that hobbits had mastered fire and established a stone tool industry on Flores. The discovery challenged views on what it meant to be human.

Brown thought at first that the hobbits had descended from *Homo erectus*, then thought to have reached Flores several hundred thousand years ago. In 1998, Morwood and colleagues got fission-track dates of more than 840,000 years for stone tools from another site, Mata Menge, in the Soa Basin of central Flores. Fission-track dating is based on the accumulation of damage in volcanic rock or glass from the fission of uranium impurities—the reaction that powers nuclear bombs and reactors. When radioactive atoms in a mineral split they leave a scar, or 'fission track', that is preserved for hundreds of thousands of years as long as the temperature of the mineral stays below a certain level reached only in violent geologic processes. The density of tracks indicates the time since the last volcanic eruption.

Dutch missionary Theodor Verhoeven had discovered the Mata Menge site in 1963, assigning the tools to *Homo erectus* and estimating their age at 750,000 years, a figure that aroused controversy.

It was argued that *H. erectus* lacked the maritime skills to get further east than Java, periodically connected to the Asian mainland by Sumatra when the sea level dropped during the ice ages. The ancient humans were stuck, along with Asian land animals, to the west of Wallace's Line. Flores, Timor and other islands to the east were separated by sea from Asia and Australia during the entire Quaternary, and carried unique species.

An Indonesian–Dutch group obtained a date of 750,000 years for Mata Menge in the mid-1990s, deploying palaeomagnetic dating. This method is based on the instability of the Earth's magnetic field, which has been flipping polarities due to changes in the motion of the liquid-iron core rotating 3,000 kilometres below us. The direction of the magnetic field in the past was recorded in rocks when they were formed, with the record stretching back 3 billion years. The last reversal of the poles, the Brunhes-Matuyama reversal, occurred 780,000 years ago.

The basal levels of the site were below the Brunhes-Matuyama reversal. But it was not until 1998, when Morwood and colleagues published results of a comprehensive study of artefacts and *Stegodon* bones deposited in the 1.3-metre-thick white ash, that Verhoeven seemed to be vindicated. The findings, published in *Nature*, suggested that *Homo erectus* was capable of building boats, a skill thought to be predicated on symbolic language.

And the hobbit was similar to, albeit smaller than, 1.8-million-year-old *Homo* skeletons from Dmanisi, in the Republic of Georgia. These skeletons, considered by some to be a *habilis/ergaster* or *habilis/erectus* intermediate, were thought to be the first hominins out of Africa. Brown reasoned that hobbits had evolved from *H. erectus* and, through long isolation on Flores, had become dwarfed. Some big animal species shrink under evolutionary pressure on islands that have little food, and too few predators or competitors to make size matter. Animal dwarfing has occurred on small islands in the Mediterranean and the Arctic, and on South Australia's Kangaroo Island—where large animals shrank and smaller ones evolved into giants. The museum of Siracusa holds a full skeleton of a Sicilian dwarf elephant. Dwarfing can also be pedomorphic, with the animals retaining less mature traits when they become adults. Some dwarf elephants had disproportionately short legs and snouts. Perhaps the hobbit was a pedomorphic dwarf *Homo erectus*.

Equally surprising was the young age of the remains. Wollongong University's Richard Roberts and Kira Westaway obtained luminescence dates of less than 30,000 years ago on quartz and feldspar grains associated with the LB1 remains. Michael Bird, Chris Turney and Keith Fifield got a radiocarbon age of 18,000 years for charcoal associated with the skeleton. They used ABOX-SC processing, previously shown to be effective at Devil's Lair, in Western Australia.

There is no direct evidence of *Homo sapiens* on Flores before about 12,000 years ago, but the modern humans' route to Australia from mainland Asia would have taken them island-hopping across Indonesia. Some say Flores would not have escaped the wave. Others, including Colin Groves, disagree. He says *Homo sapiens* might have taken the northern route through Borneo, bypassing Flores.

When, after exhaustive measurements and statistical analysis, Brown went with the *Homo* classification and named Hobbit *Homo floresiensis*, all hell broke loose. The classification put two human species in the region at the same time, toppling the multiregionalist scheme.

Under pressure from Soejono, Morwood's collaborators at ARKENAS surrendered some of Hobbit's remains to Jacob as news of the publication reverberated through the scientific community. The research team had not completed its research on the material, and argued that the move violated its intellectual property agreement as well as ethical standards.

In December 2004, Jacob, who had gone to the media with claims that Hobbit was a modern human with a brain disorder called microcephaly, returned to ARKENAS for the rest of the material.

Alan Thorne and Adelaide University palaeoanthropologist Maciej Henneberg flew in, without the discovery team's knowledge, to study the specimen. 'They seem to lack the capacity to recognise a village idiot when they see one,' Thorne famously said on Channel 9's *60 Minutes*. By then the team was finding more hobbits, suggesting that Liang Bua was an entire village of idiots. Meanwhile, Jacob and colleagues took a film crew on an expedition measuring short people living near Liang Bua, arguing that they were descendants of the hobbits.

The row over possible hobbit ailments would become chronic, with the microcephaly diagnosis arousing the most controversy. Microcephalics have brain capacities of less than 700 cubic centimetres against the average 1,400 cc for normal humans. They are usually short and intellectually impaired, and most die young. True microcephaly is largely genetic; the skulls of people with the disorder have a signature shape, with a pointed head, a small brain case and a narrow forehead that slopes back.

Thorne and Henneberg's first scientific publication on the subject was in *Before Farming*, an on-line journal which is not always peer reviewed.[1] They said Hobbit's skull closely resembled that of a 4,000-year-old microcephalic from Crete. That skull, with a capacity of about 530 cc, belonged to a 20-year-old male. Morwood and the ever-candid Brown countered in the same journal: 'This is an extremely poorly informed, and ill designed piece of "research" and could not have been published in a substantial peer reviewed journal.'

The authors have either not read the article upon which they are commenting, or have a very limited knowledge of hominin evolutionary anatomy, perhaps both.'

Dean Falk, of Florida State University, and colleagues weighed in. Falk's team, which included Brown, compared Hobbit's brain with those of modern humans, including a true microcephalic and a pygmy, along with those of australopithecines. Hobbit's brain most closely resembled that of *Homo erectus*, with *Australopithecus* the next most likely relative. It least resembled the microcephalic's. The scientists reported their results in *Science*.

A team led by the ANU's Debbie Argue compared Hobbit's skull with the Cretan specimen and data from more than 1,000 modern human skulls from around the world. The Cretan microcephalic was within the modern human range but Hobbit was way outside it.

A team led by Jacob and including Thorne and Henneberg made its case in *Proceedings of the National Academy of Sciences* in 2006, arguing that LB1 descended from a pygmy *Homo sapiens* population but also showed signs of 'a developmental abnormality, including microcephaly'.

In 2007, Falk, with a slightly different team, published further research. This time the scientists compared Hobbit's skull with those of ten normal and nine microcephalic humans of varying age, gender and ethnicity. They also compared it with a dwarf's skull. The results again ruled out microcephaly in Hobbit.

Other disorders invoked include Laron syndrome, a genetic disease affecting growth. Yet another is cretinism or congenital hypothyroidism.

During the drama over Jacob's 'borrowing' of the hobbit remains, debate in the Ausarch internet chatroom began calmly as a discussion on the ethics of the actions of Jacob, Thorne and Henneberg, and on the discovery team's right to complete its research before opening up access to the specimen. Then the exchange got a bit untidy, degenerating into a slanging match over 'power politics'. Old wounds over the Jinmium site were re-opened. The players in the Hobbit dispute were described as 'silverbacks at 20 paces'. The argument was elevated briefly with some epistemology, with references to 'Kantian reality', falsifiability and the question of whether archaeology rated as a science, but it descended again in a wave of abuse. One researcher was advised to 'take Jacob's blow on the chin … wait for Thorne's salvo, and then have your OoA [Out of Africa] troops respond'.

One subscriber, who pens scholarly theoretical texts, refined her stance when she wrote: 'It is healthy to debate (no matter your theoretical position on the nature of archaeological "science") … What has always stifled good and constructive archaeological debate, particularly in Australia, is the way in which silverbacks tear up the vegetation, thump their chests, howl and fart loudly, rather than engaging in constructive and, dare I say, theoretically sophisticated debate. Also, if I hear the "authority" of "science" invoked any further I think I will scream.'

'Are we talking about archaeology, palaeoanthropology or primato- logy?' countered Africanist David Cameron.

While some were focusing on the hobbit's bones, others turned their attention to soft tissue. Peter Schouten's dramatic illustration of a hobbit holding a spear and his quarry—a giant Flores rat—appeared on television screens and the front pages of newspapers around the world. But the portrayal of the hunter as male outraged many, who vented their spleen on Ausarch. Some even argued that the species' nickname should have been female, with 'Flo' attracting consider- able support. While 'worrying and wringing' his hands over Hobbit's disappearance, Morwood had had 'time to get a female drawn', one subscriber said.

The controversy had spilled into the wider community. In a letter to the editor of *The Weekend Australian* newspaper, Victorian Margaret Roberts wrote: 'Give over, you men. The stunning new species of dwarf human—the Flores Hobbit—was clearly a woman. So why does illustrator Peter Schouten draw her with male genitals?'

Jacob finally returned Hobbit in February 2005, but his labora- tory had damaged the fragile type specimen while making moulds, the discovery team says, a procedure it had rejected in favour of CT scans. Some of the damage was to the jawbone, hindering the Morwood team's attempt to meet a request from referees to check critical dimensions of the skull while a second paper was undergoing peer review at *Nature*.

The dispute over Hobbit's skeleton saw the Australian scientists being banned from research in Indonesia for a year until things cooled down. 'Thorne, Henneberg and Jacob were all of the "multiregionalist school" of human evolution ... Some have argued that the identification of a new human species on Flores undermines the multiregionalist position,' Morwood and Penny Van Oosterzee write in their book, *The Discovery of the Hobbit.* 'Ideology may explain why Jacob and his cohorts were so adamant that Hobbit was a modern human, and this in the face of a mounting body of evidence that said otherwise,' they write. 'This was not a scientific debate between two groups of scientists carefully weighing up the evidence in a learned and objective manner. This was about others undermining a long-term and productive cooperation between Australian and Indonesian institutions, and impeding our ability to analyse, publish and get proper credit.'

Asked to comment on why Hobbit posed a problem for the multiregionalists, Brown said: 'It [*Homo floresiensis*] is there the same time as modern humans but clearly is a different species and has none of the so-called regional features that multiregionalist people argue all hominids in that part of the planet should have. They like a single evolving lineage without branches. That's why they don't like it. That's why they've been desperately claiming pathology or some other form of problem to try to get themselves out of the big pit they're in.'[2]

As early as 2005, Hobbit's arm bones and the remains of others of her kind had been found, and Brown was having doubts about the taxonomy. In his second *Nature* paper, he reported that the hominins shared some characteristics with *Homo* but others with the much more ancient hominins, such as the famous 3.2-million-year-old *Australopithecus afarensis* Lucy from Ethiopia, or the 2.5-million-year-old *A. africanus* Taung child specimen from South Africa.

By 2007, the research team had recovered the remains of six to nine hobbits, depending on how the skeletons are reassembled, ranging in age from 12,000 to 20,000 years old. Stone tools spanned the period from 12,000 to 95,000 years and look like the 800,000-year-old artefacts from Mata Menge.

By the end of 2008, several teams had measured Hobbit, with every centimetre of her anatomy, from her wrist and foot to body parts few people had heard of, scrutinised and compared with the corresponding bits of other species, from australopithecines to every species in *Homo*. Her wrist had bones like those of a chimpanzee or *Australopithecus*, according to Matthew Tocheri, a palaeoanthropologist at Washington's Smithsonian Institution. She would not have been as good at grasping as a modern human. Her shoulder, according to Susan Larson, of the USA's Stonybrook University, is like *Homo erectus*'s.

Many were by now pointing to similarities between Hobbit and *Australopithecus*. Others were seeing different 'affinities', with *Homo habilis*, for example, which probably evolved more than 2 million years ago. Brown was still rethinking Hobbit's qualifications for inclusion in the genus *Homo*, suspecting that her lineage was much more ancient. This suggested the ancestral hominins who left their artefacts at Mata Menge were not *Homo erectus* at all. While the hobbits are not humans, they are not australopithecines either, he says. He cites mounting evidence from independent studies, some of it presented at the meeting of the American Association of Physical Anthropologists in Philadelphia in 2007, backing his view that the hominins had 'something other than *H. erectus* as the ancestor'. He says the big implication is that the colonisation of Flores, probably by accident, perhaps by colonists clinging onto vegetation powered by a tsunami, is the first evidence that pre-humans ventured out of Africa. A classification outside *Homo* would mean that the hobbits did not pose a threat to multiregionalism after all. Their colonisation of an island so far away, probably through luck, not intelligence, makes the Flores discovery even more spectacular than previously thought.

In a bizarre twist, Henneberg claimed in his popular science book, *The Hobbit Trap*, published in 2008, that one of LB1's molars showed signs of dental work, suggesting that Hobbit was a modern human alive sometime after the 1930s, when dentistry came to Flores. The claim was rebutted by Brown but got a good run in the media. Heading off the inevitable question of why he had not subjected his conclusions to peer review in the scientific literature, Henneberg said that while at least some members of the 'pathology group' had agreed with him, the team had urged him to keep the claim secret for fear of ridicule. The group also feared that 'detonating such a potential bombshell' would jeopardise publication of its manuscript, which had already been bounced by *Nature* and *Science* and was being redrafted for PNAS.

* * *

After some modern humans ventured into the tropics, others headed north-west to Europe. They were perhaps surprised to discover the land was already occupied. The Neanderthals, as the modern humans of the Holocene would call these hominins, had been the unchallenged inhabitants of Europe for more than 250,000 years, and by 150,000 years ago had extended their range to western Asia. By around 40,000 years ago, they were in central Asia and southern Siberia, but they never made it to South-east Asia.

18 Neanderthal

'Some Neanderthal supporters appear as if they are defending a human minority group which has been unfairly discriminated against ... and certainly social scientists sometimes seem to regard people like me as almost racist in the way we characterise Neanderthals,' says Chris Stringer, who did his first project on the hominins at primary school at the age of nine.[1]

Controversy over the relationship of *Homo neanderthalensis* to modern humans started when the Neanderthal was first classified as a distinct species on the basis of bones found in 1856 at Feldhofer in the Neander Valley by quarry workers digging limestone amid urban development in Düsseldorf and Köln. No-one knew then how significant the discovery, 15 years before Darwin reluctantly published *The Descent of Man*, would be in the debate on human origins. Meanwhile, Thomas Huxley's attack on problems of physical anthropology, which involved comparing an Australian Aboriginal skull with the Feldhofer specimen, shows how science would have benefited had 'Darwin's bulldog' sometimes been muzzled.

Multiregionalists and Africanists today argue relentlessly about Neanderthals. Were the stocky, archaic people with big noses, receding chins, protruding brow ridges and thick skulls attractive enough to the slender, smart *Homo sapiens* who encroached on their territory 40,000 years ago to get their genes into the modern human gene pool? Were the two groups related closely enough to produce viable offspring? The multiregionalists say the groups did mate, with the

Neanderthals being assimilated into the modern population. The Africanists say gene flow didn't happen or happened only sometimes, and the moderns replaced the Neanderthals either through a blitzkrieg or by out-competing them.

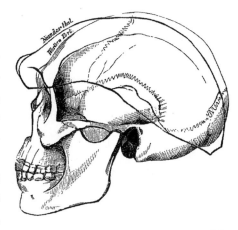

An Aboriginal skull with the contour of the Neanderthal skull, from Thomas Huxley, The Place of Man in Nature *(1863).*

The image of Neanderthals has changed dramatically since the first half of the twentieth century, when they were por-trayed as dull-witted, knuckle-dragging cave dwellers. Visual reconstruction influenced the archaeological reconstruction of Neanderthal Man, according to iconography specialist Stephanie Moser. Pictures, including one by graphic artist Kupka in the *Illustrated London News* of 1909, showing the La Chapelle-aux-Saints Neanderthal as a brute influenced generations.

According to João Zilhão,[2] an outspoken palaeoanthropologist of the University of Bristol, the hominins' bad press is a hangover from times when 'Neanderthals were used as supporting ancillary evidence in mainstream ethnological views of the racial ladder ... Today, ranking human races is no longer acceptable, but in western culture, the philosophical or religious need to place "us" at the top of the ladder of life is still very prevalent ...'

Neanderthals now have an adoring public and media pack, and their skeletons and caves are big tourist attractions. A German and Swiss team using computer-aided reconstruction created a life-sized Neanderthal model based partly on the type specimen, Feldhofer I, in time for celebrations in 2006 marking the one hundred and fiftieth anniversary of the discovery in the cave in a gorge carved out by the Düssel River. To flesh out the facial features, the team copied the soft tissue of modern Swiss people, giving the model a ruddy complexion suggesting a penchant for good food and wine. During the celebrations, the resurrected Neanderthal man posed,

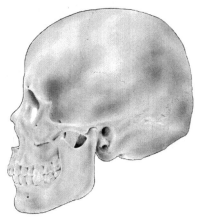
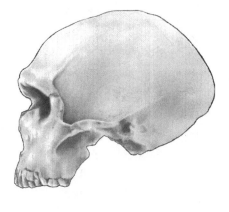

H. sapiens   *H. neanderthalensis*

with spear in hand, on a manicured lawn at the site of the Feldhofer
quarry, destroyed by miners long ago. He gazed with a mischievous
smile at Chris Stringer, Paul Mellars, Jean-Jacques Hublin and other
'Neanderthalists' who had gathered for a group photo. The nearby
museum displays reconstructions of Neanderthal families in Ice Age
environments or dressed as modern Germans in suits and ties.

Such reconstructions are popular with palaeoanthropologists
and the public alike, and genetics is filling in details. The academic
argument shifts between bones, dates, genes, signs of symbolism and
sophistication in archaeological sites, and the 'deconstruction' of
scholars convinced of Neanderthal 'otherness'.

Early *Homo* species appeared on the eastern extremity of Europe
by 1.8 million years ago as part of the out of Africa I dispersal. The
Georgian Dmanisi skulls, which some palaeoanthropologists attribute
to a new *Homo* species, *H. georgicus*, are among the oldest traces in
the region. Tantalising early glimpses come from Oldowan technology
(three cores and some flakes) dated to 1.7 to 1.3 million years old
from limestone caves of Pirro Nord, Southern Italy. The lineages of
these Early Pleistocene people, who made a living in dry open plains
and forests occupied by hippos, baboons and sabre-toothed cats,
probably went extinct. Perhaps up to a million years later, in the
Middle Pleistocene, other hominins came on the scene. They have been
assigned to a bewilderingly large number of *Homo* species—*antecessor*,

cepranensis, heidelbergensis, petraloniensis and *steinheimensis*. Many palaeoanthropologists reject some of these species designations.

The oldest known European human fossil, excluding *H. georgicus*, is a mandible associated with Oldowan technology, recently found at the Sima del Elefante cave site in Spain. It was dated to 1.2–1.1 million years old through cosmogenic aluminium-26/beryllium-10 'burial dating', the method used on *Australopithecus* remains in Sterkfontein. Long-lived radioactive isotopes, such as beryllium-10, with a half-life of 1.5 million years, and aluminium-26 (0.7 million years) are produced when cosmic rays bombard silicon grains in rocks and soils at the Earth's surface. When the grains are buried at depths of several metres, out of the reach of cosmic rays, radioisotope production stops. Radioactive decay decreases the concentration of the long-lived cosmogenic isotopes, so they work as chronometers, measuring the time since the material was shielded from cosmic rays. Accelerator mass spectrometry is used to measure the minute concentration of beryllium-10 and aluminium-26 atoms to evaluate 'burial ages' for the rocks. The method has a reach of millions of years.

The date on the Sima del Elefante site was confirmed by palaeo-magnetic and biostratigraphic dating. Spanish palaeoanthropologists associate the hominin with the expansion of eastern populations, perhaps descendants of the 1.7-million-year-old *Homo georgicus*, possibly a product of speciation at the extremity of Eurasia.

To the Africanists, these different European hominins were evolutionary dead ends, but to multiregionalists, they were variations on the *sapiens* theme. Palaeoanthropologists have been known to stop speaking to each other after arguments over the taxonomy. A study of more than 5,000 teeth from *Homo* and *Australopithecus* fossils in 2007 suggested a bigger contribution from Asian hominins in the Early and Middle Pleistocene European gene pool than accorded by out of Africa I.

From the chaos of Middle Pleistocene dispersals, one member of the human genus, *Homo heidelbergensis*, probably a descendant of the African *Homo ergaster*, evolved into *Homo neanderthalensis* in Europe—perhaps as early as 400,000 years ago—according to the Africanists. That number is based on the genetics, but archaeology puts Neanderthals on the landscape between 250,000 to 35,000 years ago. The teeth of a European *H. heidelbergensis* from near Trieste, Italy, were recently dated to almost 500,000 years. Populations of *H. heidelbergensis* that stayed in Africa evolved into the modern humans who conquered Europe about 40,000 years ago, possibly advancing along a gorge cut through the Carpathian Mountains by the Danube.

Both sides of the debate claim support in the bones for their theories. Milford Wolpoff, who sees the Neanderthals as the direct ancestors of present-day Europeans, insists their morphology is in the range of modern humans. 'Look how the skull of La Chapelle-aux-Saints and this German skull from the Middle Ages are amazingly similar, suggesting that Neanderthal features do appear in modern Europeans,' he told a conference held near the Circeo's Neanderthal site, south of Rome, as part of the Feldhofer anniversary talkfest. Forensics specialists still used Neanderthal features, such as middle facial prognathism, as diagnostic characters for Western European skulls, he said.

The so-called 'assimilation model', intermediate between the extreme forms of the out of Africa and multiregional models, allows for some interbreeding between the two species. But the question is how much, says Stringer. 'It's a question of scale—what percentage of archaic genes would we need outside of Africa to say that a strict out of Africa model was disproved?'

The question hinges on the period of overlap. Palaeoanthropologist Paul Mellars, of Cambridge University, in 2006 reviewed AMS radiocarbon dates for the dispersal of modern humans across Europe. The results suggested Neanderthal/ modern human coexistence of just 6,000 years, against previous estimates of an overlap twice as long. His calibration of the older dates has been challenged by timelords, including Chris Turney and Richard Roberts, and by some archaeologists and palaeoanthropologists, including

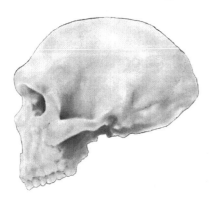

H. heidelbergensis

Alan Thorne. Wolpoff also hosed down Mellar's results. At a meeting in Bonn in 2006 forming part of the one hundred and fiftieth anniversary celebrations, he said radiocarbon dates older than 30,000 years meant 'just that—older than 30,000 years'. The amiable, larger-than-life multiregionalist, whose delivery at conferences borders on the evangelical, dumped on the discipline in which he was originally trained—physics—by questioning the capacity of AMS to date older samples. Mellars, however, was standing by his results. A later analysis went even further.

The debate gets even untidier when it turns to Neanderthal technology, art, ethics, manners and fashion.

Jean-Jacques Hublin, of the Max Planck Institute, and Mellars put a transition in stone tools from the Neanderthal's 'Mousterian' culture[4] to the more advanced Aurignacian culture[5] down to the extinction of an 'archaic' species out-competed by a 'superior' one that evolved in Africa: Neanderthals met the same fate as Pleistocene megafauna.

But Francesco d'Errico, of the University of Bordeaux, says there's little to separate modern humans from Neanderthals in subsistence strategy, stone, bone and ivory technology, and the use of colour and ornaments. The dark-skinned, tropically adapted Africans probably favoured white and red pigments for ritual body painting, while the pale-skinned Neanderthals preferred black, he says. 'The most important implication of believing that cultural modernity emerged in more than one species is that it largely eliminates the dichotomy that Western thought has traditionally detected between the natural world and human culture,' he was quoted as saying. 'The archaeology of early modern and Neanderthal populations suggests we are not the chosen people who received from God the light, the divine mandate to go forth, multiply, and eliminate their sub-human neighbours. The fact that we are the only human species left on Earth is the result of an historical accident rather than Darwinian competition.' D'Errico is uneasy about being labelled a Neanderthal-hugger, pointing out that he debunked claims that a bear bone with holes in it from a site in western Slovenia was a Neanderthal flute.[3]

Zilhão, meanwhile, believes personal ornaments, such as those found at Grotte du Renne in France, were developed independently by Neanderthals before the arrival of *Homo sapiens* from Africa. The newcomers to Europe, it seems, absorbed some cool Neanderthal stuff into their early Aurignacian culture, including tooth pendants, unknown in African *sapiens* sites. 'Neanderthals were fully human,' admits Stringer, 'but Zilhão goes too far in arguing that Neanderthals did these things first and then they taught them to modern humans.'

Some scholars say the Neanderthals buried their dead and cared for their sick and elderly. The medical history in Feldhofer I's bones indicates he had recovered from a broken arm and other injuries, perhaps hunting wounds, when he died in the Kleine Feldhofer Grotte 44 millennia ago, suggesting he was supported by his family or other members of his group.

The charge of cannibalism is a sensitive issue. Many Neanderthal bones recovered from Vindija Cave in northern Croatia were crushed, fuelling speculation that other Neanderthals had cracked them open to get at the marrow. Cannibalism is another hypothesis advanced for the Neanderthals' extinction. However, some scholars say the bone cracking could have been a ritual in a population increasingly seen as exhibiting 'symbolic behaviour'. D'Errico points to evidence that modern humans feasted on Neanderthal children, and says we've been known to eat our own kind, too.

Hopes are high that genetics will settle the dispute on Neanderthal origins. The Max Planck Institute's Svante Pääbo extracted DNA from a Neanderthal for the first time in 1997. So far, comparison of Neanderthal and *sapiens* mitochondrial DNA has failed to turn up hard evidence for Neanderthal genes circulating today.

Attention has turned to the nuclear genome with the ambitious Neanderthal Genome Project, led by Pääbo, and other studies. In 2007, an international team published research in *Science* comparing the Neanderthal and modern human form of the MC1R gene, which codes for a protein involved in the production of melanin. The modern European form of the gene suppresses melanin production, leading to red hair and pale skin. Neanderthals from El Sidrón in Spain and Monti Lessini, Italy, with ages of 43,000 and 50,000 years, respectively, had a different form of the gene, but it had the same effect on the protein, suggesting that some of them might have had red hair and fair complexions, too. In its media package coinciding with publication of the paper, *Science* issued a graphic of a red-headed Neanderthal woman alongside a modern carrot-topped man. The work had deeper implications, however. The research team, led by Carles Lalueza-Fox of the University of Barcelona, said the gene probably evolved separately in the two populations. Stringer read the result as evidence against interbreeding, because lighter skin would have conferred an advantage on the newcomers but modern human Europeans had to evolve their own version of the gene. Earlier, Holger Römpler, a member of the research team, found evidence in mammoth DNA that the extinct creatures, once hunted by the Neanderthals, could have been blonde.

Also in 2007, an international team of scientists, including Pääbo, zeroed in on the FOXP2 gene in the Neanderthal genome. Identified through a family with a congenital speech impairment, the gene is critical to speech and language in *Homo sapiens*, and the modern variant differs from its chimp counterpart by just two coding letters. Scientists found that Neanderthals had the same form of the gene as modern humans, suggesting that the common ancestor of the two also carried the gene. They reported the work, not officially part of the Neanderthal Genome Project, in *Current Biology*. But Wolpoff had a different take. 'This language gene ... seems to have got into those Neanderthal populations ... by interbreeding,' he said. 'That means there's interbreeding. That's what the multiregional hypothesis is.'[6]

Morphological studies of the Neanderthal oral cavity suggest the species had serious phonetic limitations, with capabilities comparable to those of a newborn infant, but this hypothesis remains controversial.

Some scholars saw the FOXP2 result as mounting evidence of the Neanderthals' humanity. 'Many of our supposedly modern human-derived genetic traits are shared with the Neanderthal,' said Zilhão.[7] 'This should lead us, in conjunction with the fossil evidence and the archaeological evidence, to abandon once and for all this notion that Neanderthals were a subhuman form of hominid that did not contribute in any way to the ancestry of subsequent populations.' Stringer stressed that he did not regard Neanderthals as sub-human either, but said they were 'certainly different from us and probably not our ancestors in any meaningful sense'. Neanderthal artefacts were far from testimony to 'the complexity of life and language that we have'. 'Neanderthals' social groups were probably less complex. Yes, they were talking to each other but probably they did not have the richness of language we have, regardless of whether they had the FOXP2 gene.'[8]

The man at the centre of the Neanderthal Genome Project is Vi-80, who lived in northern Croatia about 42,000 years ago. All that remains of him is a 7-centimetre-long fragment of leg bone, but his DNA confirmed his Neanderthal and male status. The project, launched in 2006 by the Max Planck Institute and 454 Life Sciences, a Connecticut-based gene sequencing company owned by Roche, was taking a 'ground up' approach as well as looking at candidate genes. It was aimed at comparing the 6 billion 'letters' in the modern human and Neanderthal genomes. By triangulating modern human DNA with that of Neanderthals and modern chimpanzees, the scientists expected to settle the debate between the two rival models of human evolution. They also hoped to see what made us different—which of the modern human genes were unique to our species. The project had the potential to reveal the differences in cognitive and behavioural traits that saw us sequencing the Neanderthal genome, not the other way around. 'The FOXP2 we did because it was such an obvious candidate,' said Pääbo. 'To our surprise, it turned out to look modern human-like. The cool thing with having the whole genome is we can look at everything. We can do an unbiased search and see what we come up with.'[9]

The task is daunting, as 98.8 per cent of the *Homo sapiens* genome is identical to the chimp's. DNA variability among *H. sapiens* corresponds on average to about 3 million bases. The difference accounts for only 36 million base pairs. Humans and chimps diverged from a common ancestor 5–6 million years ago, so about 18 million mutations occurred in our lineage. And since 95 per cent of our DNA does not code for proteins, it will be difficult to pinpoint the small number of mutations that coded for modern human morphology and behaviour.[10]

There will be even fewer genetic differences between modern humans and Neanderthals, which split from the ancestral *heidelbergensis* only 400,000 years ago.

A technological push for the project came from the 454 Life Sciences high-throughput sequencers, claimed to be up to 100 times faster than conventional instruments. Although they were designed for work on modern DNA, 454 said they made the Neanderthal Genome Project feasible by allowing about 'a quarter of a million single DNA strands from small amounts of bone to be sequenced in only about five hours by a single machine'. The technique, which is making inroads into the biopharmaceuticals industry, sidesteps the fragmentation problem besetting ancient DNA work because it uses shotgun sequencing, in which DNA is chopped into short segments as part of the process. In life, enzymes repair the DNA double helix sugar and phosphate backbone, but in death the 75 trillion cells in the body release other enzymes that break DNA down. Bacteria and fungi also degrade DNA. The resulting fragments, typically less than 200 base-pairs long, are about the size of the segments sequenced in the 454 technique.

The new 454 system was used recently to sequence, in just four months and at the cost of US$1.5 million, the 6.4 billion DNA bases of James Watson. The sequencing in 2001 of geneticist Craig Venter's genome using a previous generation machine based on the sequencing of 500–1,000-base segments cost US$100 million.

At the time of writing, Pääbo, who claimed to have come up with a work-around for the DNA damage problem and to have cleaned up an early contamination problem, expected to have a rough draft of the Neanderthal genome out by 2009.

The genetic basis of behavioural traits is not sufficiently understood to reveal whether emotions such as depression and anxiety are confined to modern humans. Did the Neanderthals, at least the French ones, have existential angst? They might have, when modern humans arrived on the scene.

In his novel *The Inheritors*, William Golding thought language was probably the significant difference between us and them. The last Neanderthals died in an idyllic Pleistocene park, perhaps in Gibraltar, probably screaming that they had been grossly misunderstood.

19 'Vampire' project

One was trained by the world's leading geneticists; the other has a mind shaped by Locke and Socrates. The geneticist and the philosopher were in 2008 at the helm of an ambitious project to collect and analyse DNA samples from 100,000 indigenous people around the world to assemble the global human family tree. The contrast between American geneticist Spencer Wells and high-profile Australian bioethicist Simon Longstaff reflects the complexity of doing population genetics research in the twenty-first century. Wells was leading the multi-million-dollar Genographic Project, while Longstaff was chairing its international advisory committee. Science had met ethics, history and politics.

The collaborative effort, privately funded by the National Geographic Society and IBM, aimed to trace the exodus of modern humans from Africa up to 200,000 years ago. Time was running out, as the 'global melting pot' was erasing the story of the past written in our genes. Collecting DNA from indigenous people was always going to be sensitive, and this time the scientists were not taking any chances. In the 1990s, a similar study, the Human Genome Diversity Project, failed after running up against opposition from indigenous leaders, especially in Australia.

When it was announced in 2005, hopes were high for the five-year program. It was well funded, with the Waitt Family Foundation, fronted by Ted Waitt, founder of the Gateway computer giant, putting up the money for the expensive fieldwork.

Australia would be critically important. 'People have been in Australia for a long, long time, much longer than Europeans have been in Europe,' Wells later said. 'It's a very important part of the overall story about humans populating the world. It has its own unique language family, very divergent languages—again, cultural evidence that people have been living there for a long time. We have virtually no samples from Australia, certainly no samples that properly represent the diversity, the cultural diversity and the genetic diversity. It is almost *terra incognito* ... We need to get more samples to be able to tell the story in more detail.'[1]

La Trobe University in Melbourne was one of 10 research institutions world wide coordinating the fieldwork component, which would increase tenfold the international database of samples from indigenous people. Other centres were in Brazil, China, France, India, Lebanon, Russia, South Africa, the UK and the US. Adelaide University's Australian Centre for Ancient DNA was to handle the work on palaeogenetics.

But in mid-2008 the results were mixed, and the architects of the project were moving to draw distinctions between it and its predecessor amid a political campaign to conflate the two. In hindsight, the earlier project had made some tactical errors.

The Human Genome Diversity Project had been the brainchild of Stanford University's Luigi Luca Cavalli-Sforza. His formidable credentials in population genetics dated back to the early 1960s, when he did pioneering work using blood groups to construct the modern human family tree and infer the pathways of human dispersal.

In the early 1990s, a group of scientists including Cavalli-Sforza and Allan Wilson, of the University of California, Berkeley, put together a proposal for a worldwide survey of variation in the human genome. Wilson was on the team that had published the famous 'mitochondrial Eve' paper in *Nature* in 1987, which claimed to trace the genetic heritage of all people on the planet back to a single African mother.

The effort would mobilise scientists from around the world, organised into regional centres. Up to 100,000 people from between 400 and 500 populations would be sampled at an estimated cost of US$30 million. Some blood samples from each population would be used to make cell lines that could be stored indefinitely as sources of DNA. The scientists would take saliva or hair samples from a larger group within each population.

The thrust was at the pure end of science. It was to be largely curiosity-driven, tracing the origins of *Homo sapiens* through the collection and synthesis of genetic, anthropological and linguistic data. It promised to answer questions about human dispersal and demographics, technological innovations, social interaction and the factors driving the development of different traits and languages.

The project architects pointed to a cultural imperative to use the 'extraordinary scientific power that has been created through the development of DNA technology to generate for the benefit of all people information about the history and evolution of our own species'. The position of Australian Aborigines in the global scheme was critical to the work, with the potential to settle the dispute over theories of human evolution.

The project would also combat racism, the coordinators claimed. 'The HGD Project will also provide the scientific data to confirm and support what is already clear from population studies—that, in biological terms, there is no such thing as a clearly defined race,' they said in a summary document. 'The results of the project are expected to undermine the popular belief that there are clearly defined races, to contribute to the elimination of racism, and to make a major contribution to the understanding of the nature of differences between individuals and between human populations.'

After years of planning workshops involving geneticists, statisticians and representatives from around the world, in 1994 the scientists won the backing of the Human Genome Organisation (HUGO), the professional society that was overseeing the medically focused Human Genome Project.

The diversity project was to have some medical objectives, too, and that proved to be one of its big political weaknesses. The genomics revolution was getting under way, with the push into one of science's last frontiers—the genetic blueprint of *Homo sapiens*, then thought to be coded in 100,000 genes.

It had been a long time coming. DNA was first identified in 1869 by Swiss doctor Friedrich Miescher in cells isolated from pus in the discarded bandages of wounded soldiers in Germany. In 1953, American biochemist James Watson and his British colleague Francis Crick, along with Rosalind Franklin and Maurice Wilkins, worked out the three-dimensional double-helix structure of DNA, but the details were not filled in until the 1970s. In 1975, British biochemist Frederick Sanger devised a method for sequencing DNA.

The international, publicly funded Human Genome Project was in full swing in the early 1990s, aiming to complete the full sequencing of the human genome by 2006. The deadline was later brought forward and work accelerated when, in the late 1990s, Celera Genomics, a US company founded by maverick scientist Craig Venter, embarked on its own sequencing project. HUGO stepped up its efforts amid fears that Celera would get there first with the grand plan and patent genes with commercial potential. It put its data in the public domain as its results came in, and both groups published draft sequences in *Nature* and *Science* in 2001.

'Molecular biology' had entered the public consciousness, and hopes were high that the field would deliver new drugs, therapies and diagnostic tests, revolutionising medicine in the new millennium. But there were fears. One centred on genetic privacy, and the risk that insurance companies and employers would discriminate against people on the grounds of genetic fitness. Another was that Big Pharma would mine the stuff of human beings, indigenous or otherwise. Those fears were fuelled in the mid-1990s by news of US patent applications, unrelated to the diversity project, on genetic material from indigenous people in Papua New Guinea and the Solomon Islands.

Unease about the new science of genomics spawned a profession. Bioethicists were on the case. Some of the practitioners were training through new university courses, while others came from medicine, law and philosophy. Still others were theologians born again as bioethicists.

Meanwhile, Australian Aborigines were getting native title claims under way after the High Court's landmark *Mabo* judgment of 1992, and the reconciliation movement was gaining momentum.

Aborigines remained the most disadvantaged people in Australia, with a life expectancy far below the national average and many of their communities disintegrating. They were still over-represented in the criminal justice system despite the shocking report of the Royal Commission into Aboriginal Deaths in Custody, handed down in 1991. The report depicted a gross dereliction of the duty of care in the system, resulting in a spate of deaths, many of them hanging suicides, among Aborigines in police cells and prisons around the country. And the release in 1997 of the Australian Government's *Bringing Them Home* report opened up old wounds. The report followed an inquiry into the forcible removal of Aboriginal children from their families, a practice common well into the 1970s.

In 1996, a bizarre, hitherto barely suppressed current had surfaced in Australian society, with the election to the Australian Parliament of a disendorsed Liberal[2] candidate riding on her credentials as the recipient of her 'fair share of life's knocks'. In her maiden speech, Pauline Hanson, a small businesswoman and single mother who later formed the One Nation Party, said: 'I won the seat of Oxley largely on an issue that has resulted in me being called a racist. That issue related to my comment that Aboriginals received more benefits than non-Aboriginals. This nation is being divided into black and white and the present system encourages this. I am fed up with being told "this is our land". Well, where the hell do I go? I was born here and so were my parents and children ... I draw the line when told I must pay and continue paying for something that happened 200 years ago. Like most Australians I worked for my land. No-one gave it to me.'

It was against this backdrop that the Italian geneticist from a big American university visited Fremantle in 1997 to try to convince Aboriginal leaders to support a project hinging on the collection of DNA from their peoples. Indigenous leaders from Australia, the USA, Canada and the Pacific islands had already made their views clear well before Luca Cavalli-Sforza's plane touched down in Australia ahead of the Human Genetics Conference.

They dubbed the Human Genome Diversity Project (HGDP) the 'vampire project', fearing it would hijack and exploit their genetic heritage, undermine their claim to their country (morally, if not legally), and shatter their world views at a time of resurgence in

their culture as they underwent a 'decolonisation of the mind'. Indigenous populations had had bad experiences with 'science' before,
and the scientists of the 1990s were carrying the legacy of pseudo-
scientists from the 1890s, like the British eugenicist Francis Galton.
In *Hereditary Genius*, Galton ranked Aborigines 'at least one grade
below the African negro', of whom he said: '... the number among
the negroes of those whom we should call half-witted men, is very
large'. (In 2007, the scientific community would be shocked when
the 'godfather of DNA', James Watson, then aged 80, was quoted
by *The Sunday Times* as saying he was 'inherently gloomy about the
prospect of Africa' because 'all our social policies are based on the
fact their intelligence is the same as ours—whereas all the testing
says not really'.)

The Human Genome Diversity Project was a public relations disaster. A statement on the urgency to get samples from indigenous
people before they became 'amalgamated' through intermarriage
was read as a push to take blood before populations went extinct.
'Ironically, at the same time as advances in technology have made it
possible to undertake a detailed study of human genome variation, the
human species is moving towards increasingly intensive amalgamation,'
the HGDP summary document said.

> Human populations have probably always been in flux but there is
> widespread interest in being able to reconstruct the dynamics of human
> populations in the time prior to known or written history ('prehistory'),
> particularly in the time before the dislocations caused by the large scale
> transoceanic/continental migrations of recent millennia.
>
> This leads to an interest in sampling those of the 'native' or 'aborig-
> inal' populations in each region—descendants of peoples present at the
> time of major incursions from other continents—who seem likely to
> have been least affected by admixture with the incoming populations.
> Study of these populations optimises the ability to reconstruct the
> ethnographic map to its state at the beginning of recorded history. Such
> ethnographically based data will improve the ability to understand the
> population dynamics of each region prior to this. In the absence of
> such sampling, the rate at which admixture and population amalgama-
> tion are taking place today is so great that in a few generations much

of the valuable information about regional prehistory will be made very difficult to reconstruct.

The then Aboriginal and Torres Strait Islander Social Justice Commissioner, Mick Dodson, said in an address to the Menzies School of Health Research in Darwin in November 1995:

> For over 500 years, indigenous peoples have been the subjects of concerted eradication programs, the objects of active abuse and violence, the victims of total neglect. And yet the scientific community is willing to preserve our genes, to preserve with science the uniqueness of our identity and the diversity of the global gene pool.
>
> But still it will take no responsibility for preserving our living cultures, even when faced with living, breathing members of our vanishing tribes, the non-indigenous world cannot see beyond itself to our inherent value as members of the human family, to the immediacy of our need.

Barbara Flick, then executive director of the Apunipima Cape York Health Council, wrote in a paper presented at the national conference of the Australian Bioethics Association in 1994:

> At this very moment the Human Genome Diversity Project is about to descend on us ... Is this the fate that science has chosen for us—that huge amounts of money be spent on taking samples from us 'endangered' species? Why aren't there the funds or resources available to us to try and turn around the early death rates of our people—to treat the children so that they don't have to live their little lives in sickness from one infectious disease to the other?

There was debate in the United Nations, too.

John Fleming, then a member of a UNESCO bioethics subcommittee and director of the Southern Cross Bioethics Institute in Adelaide, told *The Canberra Times* that the HGDP could threaten indigenous world views. 'If a particular group has an idea that human beings came into existence in a certain kind of a way ... and you then say actually the DNA indicates that your people in fact originally migrated from somewhere else ... you call into doubt the cosmogony by which their community is governed and sees itself,' said the former Anglican minister who had converted to Catholicism.[3]

Shortly before his Fremantle visit, Cavalli-Sforza said most of the opposition had come from Australia. 'Australia is perhaps the place where there has been the strongest negative response, but I presume for reasons which are totally unconnected with genetics but more connected with the fact that Aborigines have a lot of requests which they think haven't been accepted by the government,' he told *The Canberra Times*. 'I can't believe that the real problem is the Human Genome Diversity Project. The problem is of an entirely different nature.'[4]

The HGDP never gained a foothold in Australia, and the project was eventually wound back. However, it maintains a collection of cell lines at the Centre pour l'Etude du Polymorphisme Humain (HGDP-CEPH) in Paris. Non-profit research laboratories can access the cell lines at low cost. There is no Australian Aboriginal material in the samples from just over 1,000 individuals from 51 indigenous populations world wide. A recent high-resolution study using the collection confirmed the out of Africa model, and traced early modern human migrations.

Ten years after his visit to Australia, Cavalli-Sforza recounted his baptism of fire during his struggle to get the diversity project moving. The private, Canada-based Rural Advancement Foundation International had spearheaded the campaign against the project in the Americas. 'They took up arms against us, with much noise, including insults and totally absurd accusations, which they eventually partly withdrew, but [only] after having created substantial opposition, which scared the original supporters in the US Congress,' he said in an interview.[5]

In academia, most of the opposition came from a handful of cultural anthropologists, he said. 'Some charged us with racism but most anthropologists recognised the charge was clearly unjust.' It was to this group of hostile anthropologists that he was referring when, in an address to the United Nations, he said: 'Ignorance can breed fear and hate, but I have discovered that it is most dangerous when mixed with the personal political agenda of science haters.'

He sees fundamentalist Christians and postmodernists as threats to science today: 'It took three hundred years for the Catholic Church to apologise to Galileo for having forced him, a blind old man, to retract his theories by subjecting him to terrible threats. Now one anti-science movement in the USA has taken the flag—intelligent design,' he said.

'Other very recent enemies of science have especially influenced the field of anthropology,' he wrote.

> These are the so-called postmodernists. Some of them believe that science is not a search for truth, but a technology, essentially directed to fit the aims of politicians and capitalists who support it financially. Postmoderns use the emotional power of words, and their ambiguity, to undermine belief in reason. They apply the 'deconstruction' methods to repeat what the sophists did in the Greek world of the 5th and 4th centuries BCE.[6]

The later Genographic Project started well. It would ride on the reputation of the National Geographic Society and fluoresce in its warm and fuzzy image, while IBM would give it number-crunching clout in bioinformatics, the hybrid discipline in which biology meets information technology.

The project leader, Spencer Wells, a Texan, has a formidable background. Described by *National Geographic* as a child prodigy who started university at age 16, Wells did a doctorate in biology at Harvard under the supervision of the giant of evolutionary genetics, Richard Lewontin. Then, at Stanford, he worked with Cavalli-Sforza. A team of international heavyweights was on the project's advisory board, first chaired by Cavalli-Sforza and later by Simon Longstaff. They included eminent British archaeologist Colin Renfrew, renowned palaeoanthropologist Meave Leakey, and Queensland barrister Tammy Williams, then a member of the Australian Government's National Indigenous Council.

The research questions, outlined in an internal document, were big: Where were the deepest genetic lineages in Africa? Did modern humans interbreed with the descendants of earlier African emigrants, like *Homo erectus* in South-east Asia and Neanderthals in Europe? Why do the world's populations look so different?

Some of the questions were also sensitive: How do the genetic patterns in Australia correlate with the Aboriginal songlines—their own oral histories? Can we use genetics to trace the spread of Polynesians and Micronesians from island to island in the Pacific?

How many waves of migration were there into the Americas, and was one of them along the coast? Could Europeans have migrated to the Americas thousands of years ago? These questions touched on the sensitive case of Kennewick Man.

The project put in place a strict protocol covering privacy and informed consent. Before regional centres could collect samples, they had to get ethics committee approval from their host institutions, usually universities, and conduct outreach programs to brief interested indigenous communities on the project, a recognition of community as well as individual 'ownership' of DNA. Ultimately, the decision on whether to participate would come down to the local level, the project stated. 'While we also seek advice from regional, national or international representatives of indigenous people, we must ultimately defer to local indigenous and traditional people.' Finally, donors had to sign a consent form, but would have the right to withdraw consent.

The project sidestepped the problem of gene patenting. It would not collect medical data on donors, investigate genes of known medical relevance or conduct medical research, although it would record linguistic and cultural anthropological information. The non-profit organisation would not patent genes; scientists would publish their analyses and release the sequences into the public domain, posting them on a database for use by the research community.

In contrast to the earlier Human Genome Diversity Project, the Genographic Project would not make transformed cell lines, because many indigenous people were averse to the idea of their tissue living on after their death. And John Mitchell, project leader in Australia, developed a mouthwash for collecting DNA, obviating the need for blood samples, a sampling practice abhorrent to many Australian Aborigines.

The aim was to get 100 samples, if possible, from each population. The work was focusing initially on mitochondrial DNA, with the possibility of extending the research to the Y and X chromosomes.

The scientists were zeroing in on genetic markers, which define haplogroups, the branches in the tree of humankind that grow in the four dimensions of space and time. Markers show where and when haplogroup clans dispersed, with the timing revealed by the amount of genetic diversity between populations. The longer the separation, the greater the genetic difference as mutations accumulate in the separate populations. Scientists use markers to trace the common ancestor of populations and assemble the most parsimonious family tree. According to the Genographic Project, the descendants of the 200,000-year-old Mitochondrial Eve split into two groups, called L0 and L1, that lived in Africa for tens of thousand of years. Haplogroup L0 probably originated in East Africa 100,000 years ago and migrated through a wide part of sub-Saharan Africa. Today people with this marker form part of the populations in central Africa and southern Africa.

The L3 haplogroup first appeared 80,000 years ago and was the first large group to leave Africa, moving north. Today the L3 marker is found in populations in North Africa and the Middle East. Haplogroup L3 split into groups, including haplogroup M, which first appeared 60,000 years ago and crossed the waters of the aptly named Bab-el-Mandeb (meaning gate of tears) strait at the start of the coastal migration east to South-east Asia and Australia.

The Y chromosome tells a similar story and is one of the most closely studied parts of the genome in population genetics. The oldest Y chromosome lineage is defined by marker M91 that first appeared 55,000 years ago. One of the branches of this group led to a man carrying marker M130, haplogroup C, whose ancestors formed part of the first wave of migration from Africa. Many millennia later, between 15,000 and 20,000 years ago, men with haplogroup Q, carrying the marker M242, made the trip across Siberia and Beringia to the New World.

Meanwhile, under another component of the Genographic Project— the public participation and awareness campaign—anyone could buy a kit for $US100 and send a cheek swab sample in to trace their own roots. The data would be stored anonymously, and much of the money raised would go to a 'legacy fund' to promote the preservation of indigenous languages and cultures.

The project made the same appeal as its predecessor for support on the grounds of urgency, but worded it better.

In a shrinking world, mixing populations are scrambling genetic signals. The key to this puzzle is acquiring genetic samples from the world's remaining indigenous and traditional peoples whose ethnic and genetic identities are isolated. But such distinct peoples, languages and cultures are quickly vanishing into a 21st century global melting pot.

The Genographic Project will provide a snapshot of genetic human variation before we lose the cultural context necessary to make sense of the genetic data. Ultimately, we hope that the findings from the project will underscore how closely related we are to one another as part of the extended human family.

Genomics was starting to become passé. The human genome had been sequenced and scientists were stunned to find that only about 20,000 genes carried the blueprint of *Homo sapiens*. The nematode worm had a very respectable 18,000 genes and the fruit fly 13,000. The genomes of several other species, including the mouse and chimp, had been nailed, giving clues on interpreting the human genome. The first marsupial genome completed was not of a species from the 'land of the marsupials', however, but the American opossum. Scientists were also surprised by confirmation that the so-called 'junk DNA', which accounts for 98 per cent of the genome but does not code for proteins, had a big role to play.

Genomics was paying medical dividends, with the identification of several disease-causing genes clearing the way for 'rational drug design' to combat them. But gene therapy had not lived up to early expectations, and fears about genetic privacy and gene patenting remained. Opposition to large-scale population genetics remained strong among indigenous groups concerned about the weakening of land rights claims and the collision of cosmologies. Their campaign against what they labelled 'the vampire project revamped' was a re-run of the one against the HGDP, with the dispute playing out in the media and the United Nations.

The Indigenous Peoples Council on Biocolonialism, a US-based group describing itself as 'an indigenous organisation that addresses issues of biopiracy', said in a 2005 statement: 'It's interesting how in the past racist scientists, such as those in the eugenics movement, did studies asserting that we are biologically inferior to them, and now they are saying their research will show that we're all related to each other and share common origins. Both ventures are based on racist science and produce invalid, yet damaging conclusions about indigenous cultures.' On cosmology, it said: 'Our creation stories and languages carry information about our genealogy and ancestors. We don't need genetic testing to tell us where we come from.' On ancient DNA research, the statement added: 'We will not stand by while our ancestors are desecrated in the name of scientific discovery.'

The sentiments resonate with those of some scholars commenting on archaeology. '... Some aspects of archaeology may well remain irrelevant to Aboriginal people ... who have their own perfectly valid view of how the world is constructed,' wrote Heather Burke, Christine Lovell-Jones and Claire Smith.

> And why should Aboriginal people be badgered to accept a 'scientific' interpretation of their own cultural material? ... a multiplicity of indigenous views only presents a problem if it is assumed that there is an ultimate need for a single correct position to emerge. In our view, consensus is often nothing more than a contemporary variant of habitual European appropriation.

Laurajane Smith claimed:

> Debates which emphasise epistemology or refutation only ensure that archaeological knowledge is assessed for its legitimacy within a certain Western intellectual framework. This framework is becoming increasingly irrelevant within a politicised archaeology, and fails to recognise the legitimacy of either indigenous or other forms of knowledge.

Simon Longstaff says that prior, not first, occupation is what counts in land rights claims. 'The strongest arguments that are ever made by indigenous people in their political, social and economic claims don't rest most strongly on whether you're the first, but the fact that you are the prior occupants of a particular territory, and then there is a whole lot of moral arguments about promises that were made and justifiable expectations. From an indigenous Australian perspective, there's nothing that any genetics study can do that will take away from the fact that they were the prior occupants to any European settlers.

'There's no doubt that there were established populations of North American native people prior to European settlement,' he adds. 'But suppose the claims about Kennewick Man are true—that there were others there before—I don't see that it necessarily follows that that eliminates the claims by indigenous people in North America.'

From a moral perspective, even if the ancestors of indigenous Americans replaced established populations, it would not exculpate European settlers doing the same thing.

He says a bigger problem is whether the scientific narrative could sit alongside the indigenous one. The wider indigenous community already accommodates contradictory narratives, the philosopher argues.

'Such narratives are sometimes mutually exclusive, in that it can't be true that one group emerged as the navel of the world and creation emerged out of them and their particular relationship to the land, and also true of another community making exactly the same claim but on the other side of the world. They can't both be right, but they don't need to be.'[7]

The UN action has had no direct effect. Says Wells: 'We have not halted the project ... We have entered dialogues with a lot of indigenous advocacy groups. It would be nice if we could get endorsement from every indigenous group but in some cases we won't and people have the right not to participate and even say they don't like the project.'

And advocacy groups could not 'purport to speak for every indigenous person'. 'There may be certain similarities among indigenous groups, and, in many cases, they're marginalised groups. They're often the poorest of the poor in very poor places. I completely understand the concerns that indigenous people might have about this kind of work, given the history of what's gone on in the scientific community. That is why we are so careful to do things the way we do them. In addition to allowing indigenous participation and allowing indigenous people to find out about their own history ... we want to give something tangible back with the legacy grants.'

Asked about the Indigenous People's Council on Biocolonialism, he said: 'Can one entity, one group, in this case, a relatively small group of individuals, purport to speak for every individual in a region or around the world? I would argue, No!'[8]

Says Longstaff: 'In indigenous cultures ... at least in Australia, the actual authority for any group is not vested in any regional, national or international bodies but instead always lives on the ground with a local community and its attachment to a particular piece of land ... From what I've been able to see, this is pretty much the pattern for indigenous people around the world.'

Both Wells and Longstaff contrast the stance of local communities with that of politicised players. 'The great thing about being out in the field with indigenous communities—when you get outside of the political sphere in New York, let's say, when you get out on the ground and start talking to them, they get really excited about this stuff,' says Wells. 'Indigenous people have a very strong sense of their history. In many cases, that's all they have. They may have the land they live off and their history and that's about it. They tend to get very excited about the idea that they're carrying a piece of their ancestors inside of them, and you can learn about that through the study of DNA, and that also shows how they're connected to other people around the world. We try to do this in the most sensitive way possible. It's not a question of science trying to replace traditional knowledge. We want to add to their sense of self. They have a strong sense of being connected to the place where they live. In addition, there's a deeper history. It's not that we're trying to say you don't have this close connection.'

By mid-2007, 25,000 indigenous and traditional people in regions including Europe, the Middle East, sub-Saharan Africa and East Asia had been sampled. However, sampling had not begun in Australia, where the social and political landscape was little changed from the time of Cavalli-Sforza's 1997 visit. It was the fortieth anniversary of a referendum in which most Australians had voted to allow Aborigines to be counted in the census and for the Australian Government to legislate for their wellbeing. The occasion refocused attention on indigenous communities, and Aboriginal affairs returned to the political agenda amid the frantic campaigning of an election year that would eventually see the Australian Labor Party returned to power after a decade in opposition, and a long-awaited apology to Aborigines by the new prime minister.

Aboriginal life expectancy was still below the national average. Native title had failed to deliver on its promises, and Aborigines were still dying in custody. A shocking Australian Government report revealed widespread child abuse in troubled Aboriginal communities, prompting the government to mount an emergency response and send teams of doctors, police and military personnel into the Northern Territory, a move that aroused controversy and divided Aboriginal leaders.

Says Wells: 'There are some obvious issues to deal with in Australia—the history of colonial exploitation ... There's a long history of animosity ... We're taking a very slow, careful, measured approach ... It's more of a challenge to sample in Australia than it is in a lot of these other places.'

This time, however, there was apparently no organised opposition in Australia.

Asked about the argument over human origins, Wells said: '[The debate] does get quite heated, but ... if you make a fair appraisal of the evidence out there on the genetic side, the archaeological evidence, the palaeoanthropological evidence, it all very clearly tells us we came out of Africa very recently and there's no evidence at the moment for interbreeding with other species that might have been out there. We're all sons and daughters of Africa ... As scientists we're open to changing that story and it would be fascinating if we found evidence for admixture with *erectus* in South-east Asia or evidence for an out of Australia model of human population expansion, but at the moment there is no good evidence for that.'

***

A study led by Doron Behar, of the Rambam Medical Centre, Haifa, and Spencer Wells used 624 complete mtDNA genomes from sub-Saharan Africans to reconstruct the early story of modern humans, a story lacking detail from the palaeoanthropological record. The work, published in the *American Journal of Human Genetics* in

2008, suggested that few of the maternal lines that arose in the first 100 millennia of *Homo sapiens* survived: our species went through a population bottleneck and came close to extinction. For tens of millennia after the start of the exodus, more than 40 matrilines grew in isolation in Africa, but only two left the continent. Most of the groups reunited during the Late Stone Age revolution of about 40,000 years ago. Only the Khoisan remained isolated, until a few centuries ago when their territory was invaded by other African tribes from the north and by Europeans from the south.

By late 2008, the Genographic Project had collected samples from more than 41,000 indigenous people world wide. Sampling had just begun in Australia, with the support of Aborigines in Brisbane.

In 2007, British geneticist Paul Brotherton and colleagues, including the Australian Centre for Ancient DNA's Cooper, came up with a method for tackling the problem of contamination in ancient DNA research. SPEX, or single primer extension, is an alternative to the polymerase chain reaction amplification method—PCR. Conventional PCR discriminates in favour of contaminants because modern DNA is usually in longer, undamaged segments and is easier to copy. With SPEX, a greater proportion of the really ancient stuff ends up in the amplification products. ACAD is to use the technique in the Genographic Project.

However, the project's progress on genetics research on modern Aboriginal DNA was not matched by work on ancient Australian skeletons, which were still out of bounds. And at the close of the first decade of the twenty-first century, the repatriation of human remains was gaining momentum. Many of the skeletons would be lost to science at a time when science finally had advanced tools to study them.

20 Back to country

In late 2006, the board of trustees of Britain's Natural History Museum met in London to discuss the fate of Aboriginal skeletons about to be lost to science. The museum was due to hand the remains of 17 individuals over to the Tasmanian Aboriginal Centre (TAC), which planned to cremate them. The board approved further tests on the remains ahead of the handover, a move that incensed TAC and sparked action in the British High Court.

The Natural History Museum was among several overseas institutions returning remains following a 30-year campaign by Aboriginal leaders in parliaments, the United Nations and the media.

The controversy had its roots deep in Australia's colonial history. The biological revolution ignited by Charles Darwin coincided with a quickening of the dispossession of Aborigines. Dismissed in the nineteenth century as primitive and locked in the Stone Age, they were viewed as interesting specimens by natural historians in the colonies and Europe. Burial grounds were raided, probably illegally under British law. Some remains were held in Australian institutions. By the 1890s, the Adelaide School of Medicine and the South Australian Museum had amassed big collections of skeletons, many from the Northern Territory. The Australian Museum in Sydney, Australia's oldest museum, was an enthusiastic collector, too, under the curatorship of Edward Ramsay, and the Museum of Victoria also housed big collections.

Other skeletons were traded on the international market, with many of them put on display. The remains of at least 3,000 individuals went to museums. At least 70 institutions in 21 countries had Aboriginal remains in their collections, according to figures compiled by the now-defunct Aboriginal and Torres Strait Islander Commission (ATSIC). Some of the remains were of people so recently alive that their names were known, and there was a premium on Tasmanian material. Truganini, whose skeleton was dug up two years after her death in 1876 and given to the Royal Society of Tasmania, was cremated in 1976. And a Tasmanian skull solicited from George Robinson, later the Protector of Aborigines, by Governor and Lady Franklin in the late 1830s found its way into Britain's Natural History Museum.

Some of the remains, perhaps of victims of frontier conflict, had bullet holes. Collecting continued into the twentieth century. Engineer George Murray Black exhumed more than 1,600 skeletons from burial mounds along the Murray in the 1930s and 1940s, sending them to the former Institute of Anatomy in Canberra and the University of Melbourne to form the 'Murray Black collection'.

The repatriation push gained pace in the late 1980s, with Aboriginal leaders making their case to museums, universities, parliaments and the United Nations. ATSIC led the campaign, but Aboriginal leaders from Tasmania fought hard, too.

Some academics have weighed in, some of them working through the World Archaeological Congress (WAC). The WAC split off from the learned society, the International Union of Pre- and Proto-historic Sciences, in 1986 amid furore over the exclusion of South African archaeologists from the IUPPS's eleventh congress in Southampton, UK, during the international campaign against apartheid. British archaeologist, the late Peter Ucko, a former director of the Australian Institute of Aboriginal Studies (the predecessor of the Australian Institute of Aboriginal and Torres Strait Islander Studies), pushed ahead with an alternative conference when talks between IUPPS factions broke down. The move projected the WAC from the more conservative body. Australian archaeologist Claire Smith, of Flinders University, was elected WAC president in 2003.

The battle on Australian soil has been won. Among the first remains held by Australian institutions to go back to an Aboriginal community were the precious Victorian Kow Swamp skeletons, returned to the

Echuca Aboriginal Cooperative Society by the Museum of Victoria in 1990. The museum does not know the fate of the skeletons. 'It is believed that they were reburied,' a spokeswoman said. 'The museum had no involvement with the remains after they left the collection. It is usual for remains to be buried after repatriation. Aboriginal people believe that if remains are not returned to country, the person's spirit will wander endlessly.'[1] (Some Aboriginal groups believe that the time of death must be known for the peace of the spirit. In the mid-1990s, an Aboriginal group from near Sydney took an ancestor skull, carefully wrapped in a kangaroo skin, to the Lucas Heights laboratories of the Australian Nuclear Science and Technology Organisation, seeking an AMS radiocarbon age. Chemist Quan Hua carefully collected a sample using a dentist's drill under the apprehensive eye of the clients as they accompanied the operation with the hypnotic sound of the didgeridoo.)

Some of the Kow Swamp skeletons are fossilised, and some are very robust, the Natural History Museum's Chris Stringer says. 'They document variation beyond anything we see today. Modern humans today are a biased sample of what people were like 5,000 or 10,000 years ago—we've had so much change in this time with population movements, dietary changes that have driven a reduction in the size of the face and teeth. This material tells us about human variation in the past.'

Amid the repatriation campaign, scientists have found themselves in an asymmetric battle, burdened with the image of the unethical, amateur collectors of the past as they face the political players of the present.

Many scientists and archaeologists, including John Mulvaney, contend that ancient remains are the common heritage of humanity. Many raise questions about the relationship between ancient remains and the communities receiving them, and point out that no-one knows the funerary rites of the ancients. Research on skeletons could shed light on the life expectancy, health, cultural practices, diet and mobility of past populations. And Australian remains could help settle debates in human evolution. Mulvaney, who favours 'keeping places', has described the reburial of ancient remains as 'reverse cultural imperialism' that obscures blacks' and whites' views of the past. 'Total destruction today, or the withholding of permission to undertake field research, not only negates the spirit of open inquiry for all people, it denies that freedom for present and future Aboriginal scholars,' he wrote following a campaign with Alan Thorne to save the Kow Swamp collection.

Many scientists support the return of younger skeletons, and some of the most hard-nosed scientists have helped the National Museum of Australia provenance remains that lack documentation. At the request of Aboriginal communities, they backed up the museum's repatriation team as it provenanced remains due for repatriation from the University of Edinburgh and the former Institute of Anatomy in Canberra. Most of the skeletons were less than 10,000 years old, with some dating from historical times, and many had been taken apart for comparative anatomical studies.

Michael Westaway, then a physical anthropologist on the repatriation team, and colleagues used bone pathology, anatomical measurements and statistics to piece remains together. Timelord Rainer Grün dated a skull from central Queensland with a bullet wound. And geochemist Wolfgang Müller provenanced the remains of a woman from Victoria by comparing oxygen and strontium isotopes in her bones with water and soil samples. Müller, now at the Royal Holloway University of London, had previously pinpointed the birthplace of the Neolithic European, Ötzi the Iceman.[2] The Aboriginal woman was returned to Port Fairy, where her clan wrapped her in bark and reburied her. In contrast, Ötzi is usually on display in a museum at Bolzano in the Italian Alps. He has been studied closely by schoolchildren as well as scientists.

Many Aborigines draw no distinction between ancient and recent skeletons, and afford no weight to how closely living populations are related to the remains, for which they feel a strong sense of responsibility. Some are shocked to learn of the skeletons and the circumstances of their removal, but have differed on how they have dealt with the remains. Some have wanted them returned 'to country' and have reburied them in moving ceremonies, sometimes these days relearnt from ethnographic accounts of funerals from the contact period. Others have favoured 'keeping places' or have asked museums to hold them, options that leave the way open to future research. Still others don't want them back immediately, because they have more pressing, social, problems to deal with or because they fear handling the remains. Historian Paul Turnbull, an expert in the history of the procurement of and research on indigenous Australian remains, has said previously that some museums had come under pressure from mainstream politicians to get the remains 'out the door', but communities receiving them were grappling with other problems, like native title claims or education or alcohol problems. One Queensland community had blamed deaths in custody on repatriation.[3]

Traditional Aboriginal societies do not have a designated occupation to handle the dead. Relatives preside over elaborate and lengthy rituals. The ceremony is designed to clear the spirit from living kin, the late anthropologist Ken Maddock of Macquarie University once said. Repatriation has created new problems for some communities.[4]

Repatriation was escalated to the highest level of government in 2000 when then Prime Minister John Howard discussed it with his British counterpart, Tony Blair. The meeting, in London, followed talks between the institutions and Australian ministers and officials in late 1999 and early 2000 in the lead-up to Australia's 2001 centenary of federation celebrations. In a joint media statement, Howard and Blair vowed to step up efforts to send Australian remains held in British institutions home.

In 2001, the Blair government set up a Working Group on Human Remains in Museum Collections.

In a submission to the committee, TAC called for the unconditional and mandatory repatriation of all Aboriginal remains. 'The spirits of our dead await the return to the traditional areas of the body parts, so the traditional ceremony that takes body and spirit back to the land can take place,' it said. 'Until then, the spirit remains in a stage of torment, lost from the human individual of which it was always a part ... No longer is indigenous culture the playground of scientists.'

In its report, handed down in 2003, the working group, chaired by barrister Norman Palmer, said more than two-thirds of requests for repatriation had come from Australian Aborigines and Maori, with most of the others coming from Native Americans. 'For these peoples, regaining control of human remains has become part of a larger set of efforts towards asserting cultural vitality and self determination,' it said. The report recommended legislative reform to allow museums to repatriate. The committee did not explicitly mandate the return of remains, but its recommendations on consent amounted to de facto compulsory repatriation. It called for a ban on retention of or research on remains known or suspected to have been obtained without the prior consent of the dead person or his or her close family. It also wanted a ban on ongoing research on remains unless the institution had obtained the consent of 'close family or genealogical descendants of the deceased person' or, if such descendants could not be identified, 'those who have within the deceased person's own religion or culture a status or responsibility comparable to that of close family or direct genealogical descendants'.

In a statement of dissent, Neil Chalmers, director of the Natural History Museum, suggested that the recommendations amounted to mandatory repatriation by the back door. 'This recommendation would, if implemented, introduce a regime of widespread mandatory return,' said Chalmers, the only scientist on the 10-member committee. '... These recommendations propose that consent from genealogical descendants or their community surrogates must have total priority over any other considerations irrespective of the age, or certainty of identity of the human remains in question, or the distance of relationship between the deceased and the claimants. They take no account of the public value of research based upon the human remains.'

Writing in the British journal *Antiquity*, Laurajane Smith welcomed the working group's report as a 'step toward acknowledging the legitimacy of claims of descent made outside of the temporal and genealogical criteria that often underpin British and wider Western conceptualisation of kinship and descent'. Of critics of the recommendations she said: 'Much of this rehearses the dire warnings about the "end of science" and the assault on the "academic freedom" of archaeological research that were made in the Australian, New Zealand and North American media and archaeological literature in the 1980s and 1990s. These criticisms, underpinned as they so often are by a discourse laden with assumptions about the unassailable "truths", "objectivity", "rights" and "universal relevance" of science, including archaeological science, miss the point ... [F]or many indigenous people the issue of depth of time simply does not apply, and the age of the remains does not necessarily temper the intensity of the ancestral link,' she noted. 'In some, but not all, Indigenous cultures identifying direct biological linkages may be irrelevant, and entirely different criteria are used for identifying ancestral/descendant links.'

Norman Levitt, of Rutgers University, challenges the idea of cultural continuity over tens of millennia. 'The history of humanity, especially of small social groups, is one of continual migration, disintegration, fission and amalgamation,' he says. 'Neither language nor culture, and certainly not geographic location, remains stable for more than a small fraction of that time.'[5]

In 2003, the WAC issued a press release urging indigenous Australians to step up their repatriation campaign. 'So far they have concentrated on Britain, and they have had real successes there, but there are also human remains in American institutions, most notably the Smithsonian Institution,' said Claire Smith. 'The next focus of international repatriation efforts will have to be the US.' On the dispute with the Natural History Museum, she said 'the vast majority' of Australian archaeologists would 'strongly support the repatriation of this material'. 'Let me be quite clear on this. It is only a matter of time. These ancestral remains will be returned to the care of Indigenous Australians.'

The worst fears about the Palmer report were not realised in UK government guidelines later drawn up. However, political pressure, if not legal pressure, remained heavy, and by 2006, six UK museums, including the British Museum, had committed to repatriate. And the *Human Tissue Act* was amended to remove legal obstacles to the removal of material from their collections. Between 1990 and June 2007, the remains of more than 1,300 individuals had been returned to Australia, most of them from the UK, but also from Sweden and the USA, according to a spokesman for the then federal department of Families, Community Services and Indigenous Affairs, which was responsible for the return of remains from overseas.[6] In 2007, the government was working to get comprehensive inventories of Australian remains in international holdings while negotiating repatriation from institutions in the UK, the USA and Europe. Provenanced remains were returned to communities, which decided on what to do with them. Remains of unknown origin went to the National Museum of Australia.

The issue came to a head earlier that year over the Natural History Museum's plans to collect more data from the Tasmanian remains destined for repatriation in line with a request in 2005 from the Australian Government for a handback of all Australian material. Repatriation would halt research on the remains, dating from the early nineteenth to the early twentieth centuries, on subjects ranging from evolution through medicine to forensics. Scientists from around the world, including Australian researchers, had studied the unique collection, and the work had generated many high-impact papers, belying claims that remains have been languishing in overseas collections. The museum wanted to collect more data because 'it had been indicated that the Tasmanian remains would be destroyed on return', according to minutes of a board of trustees meeting.

Barrister and veteran activist Michael Mansell reportedly labelled the tests 'mutilation'. The Tasmanian Aboriginal Centre got an interim injunction on the work, and won federal government funding for its legal action. The court lifted the injunction after the museum undertook to drop plans for DNA and stable isotope analysis, limiting its tests to CT scanning, photography and anatomical measurements. The dispute was later settled by mediation, and the remains returned.

The Natural History Museum's Chris Stringer laments the loss to science of the Tasmanian collection, even though the remains were probably less than 200 years old.

DNA analysis of the 100- to 150-year-old remains of Andamanese people, who are central to the debate on human evolution, challenged claims that the islanders, isolated for tens of thousands of years, were a relict population descending directly from the first modern humans out of Africa. 'Genetically, it turns out they are very much Asian,' Stringer says.[7] 'These are Asian people, and their skeletons contain

Asian-related DNA lineages, but ones that are so far unknown from living populations. In the 20 individuals we've sampled, there are DNA types which would otherwise not be known to science.

'In the Tasmanian material, there could be DNA lineages that are unknown in Australia otherwise. If that's destroyed, that's lost information, not only for us as scientists but for Tasmanian people as well in terms of their history, so even recent material is potentially important.' Further analysis of the skeletons could have revealed the Tasmanians' closest mainland relatives, their population size and the number of waves of migration before Tasmania was cut off from the mainland 14,000 years ago. It could also have placed the population in the global scheme of human diversity, and helped solve the mystery of why the Tasmanians stopped eating fish.

At the time of writing, a question mark hung over the museum's collection of mainland Australian remains—ranging from single teeth or bones to full skeletons—of about 200 individuals. Its holding of 100 hair samples collected from the living was also in the balance.

The Australian material has the potential to settle the debate over whether Neanderthals were modern humans or a separate species. 'You can only tell whether they're a separate species if you have a true measure of what *Homo sapiens* is—and whether Neanderthals are outside the range of modern humans. If we lose all that material, we lose the ability to document the full geographical range of modern humans. That's a loss for the whole of humanity. It's like you are taking a whole continent out of the history of modern humans.'

At the time of writing, fallout from the Kennewick Man dispute continued, with moves to water down the requirement of Native Americans to prove cultural affiliation with remains in order to claim them under the Native American Graves Protection and Repatriation Act. Norman Levitt described the legislation as 'a well-intentioned piece of legislation that has grown into a monster of anti-scientific bias ... initially devised to make amends for generations of arrogance and cultural insensitivity on the part of anthropologists, archaeologists, ethnographers, pot-hunters, and collectors of pre-Columbian art'.[8]

The moves alarmed scientists, who feared they would essentially halt research on ancient remains. The WAC was active in a campaign lasting several years. In 2004, the organisation issued a statement voicing support for a proposed amendment to the Act which would 'recognise as Native American all human remains found in the United States that date prior to the documented arrival of European explorers'. Alan Schneider, who acted for the scientists in the Kennewick Man case, was quoted by *Nature* as saying of one proposal: 'Even Adam's and Eve's remains, if found in this country, would be subject to claims by tribes.'

Fossil hominins from Africa are scattered around the world. In a paper submitted to the Science for Cultural Heritage Conference in Trieste in 2006, Phillip Tobias called for repatriation of the fossils to their countries of origin. However, he said a different set of issues may arise in cases involving younger human remains more likely to be the subject of repatriation requests. 'Such claims may raise questions on the credibility or authenticity of the claimed relationship between living populations and skeletal remains,' he said.

However, in contrast to their counterparts in Australia and the Americas, indigenous Africans generally support research on human remains. Repatriation of remains, even fairly recent ones, does not usually threaten research, with the skeletons usually going back to museums. In general, Africans have won their liberation struggles, and are proud of the science, which puts the roots of humankind in Africa.

'In southern Africa, there has recently arisen a series of movements, isolated from one another but perhaps not unconnected, for the return of some remains either to South Africa, or within South Africa to the area of discovery,' Tobias says.[9]

'The people living close to the limeworks of Taung have been pressing very insistently that we "return" the Taung child skull to them. [However] there is no suitable institution, museum or university, close to Taung, nor are there personnel there available who would be regarded as qualified curators. Moreover, the area is 500 kilometres away from Johannesburg and Pretoria and this would make access to the skull extremely difficult. Moreover, it is impossible to claim that the species represented by the Taung skull [*Australopithecus africanus*] was ancestral to the present-day people living in that area, 2.5 million years later! So there are many strong arguments for the skull not to be sent to the

Taung area although we are trying to persuade the people there that casts and our expertise would be made available to them to help in the setting up of any little local museum in or close to the limeworks from which the skull was obtained. Although Wits University houses and curates the skull, I believe it would need a government decision in this respect. The government is very supportive of our research and would tread very carefully in handling this request.'

However, the skeletons of some Griqua people, one of the recent South African indigenous groups, which Tobias and his students exhumed from Campbell in Griqualand in the 1960s, were returned and reburied in 2007.

Meanwhile, the French government recently returned the remains of Khoe Khoe lady, Sarah Bartman, from the Musee de l'Homme to South Africa, and they were buried at Hankie, near her birthplace. Khoisan people assembled at Capetown airport to receive her remains. Tobias, who has been an anti-apartheid activist all his life, said in the commemoration speech at the ceremony that she is a 'key icon' in the history of the infringement of the rights of men and women.

Tobias saw her remains for the first time in a public display in the Musee de l'Homme in 1955. 'Apart from her well-preserved, articulated and mounted skeleton, as far as can be ascertained only two parts of her soft tissues had been preserved: her brain and her external genitalia' he wrote. 'They were kept in embalming fluid in sealed glass jars.'[10]

In 1810, at the age of 21, she was taken from Cape Town to Europe as an anthropological curiosity and displayed under the name of the 'Hottentot Venus'. 'It [Sarah Bartman's story] incorporates attitudes of racial superiority and inferiority, and of a prurient sexism, for the London and Paris audiences who paid to see her were inordinately curious about her sexual features—as were the French scientists, especially Georges Cuvier.'[11] On her death at the age of 27, Sarah Bartman was denied a burial, and her remains fell into the hands of museum curators. Her body parts had been displayed to the public until 1994, according to the Griqua National Conference.

'As regards more recent remains such as those of the Khoisan peoples exhumed in the eighteenth, nineteenth and twentieth centuries, as well as the dissection and teaching specimens (which include all races) in the departments of anatomy in the country's medical and dental schools—there has been no concerted attempt for these remains to be removed for burial, unlike what happened in Australia,' Tobias says.

***

In the first decade of the twenty-first century, the Australian Pleistocene remained shrouded in darkness. Little pure research was being conducted on Australian Aboriginal skeletons or archaeological sites. The Australian Research Council's funding of Quaternary research through its Discovery Grants scheme dropped to $6.9 million in the 2008 round (for projects starting in 2009) from about $10.5 million in each of the previous two years, according to an analysis by Simon Haberle, of the Australian National University. Of the 16 prehistoric archaeology grants in 2008, just under a third were for research based in Australia or New Guinea, continuing a long-term trend. Many scientists and archaeologists were heading offshore, to South-east Asia and Africa, for example.

The obstacles to research in Australia and the Americas come at a time of big strides in the techniques for studying the past. Advances in dating, and DNA and stable isotope analysis, are expanding the information that can be extracted from bones and artefacts. And the physical anthropologists are rising to the challenge mounted by the geneticists, with new x-ray technology that enables them to make out minute detail in bone and teeth to track evolution. Synchrotron radiation—'bright' x-rays produced by an electron beam accelerated in a ring at near light speeds—allows scientists to generate three-dimensional images of bones and teeth at thousands of times the resolution of ordinary x-ray images, and without destroying the specimen. The technique promises to unlock the secrets of developmental biology for Neanderthals, *Homo erectus* and other hominins. Using synchrotrons, scientists recently discovered that Neanderthal tooth enamel was thinner than that of modern humans, supporting the hypothesis of a faster childhood development for Neanderthals. And by comparing strontium isotopes in the molar of a 40,000-year-old Neanderthal from Lakonis, Greece, with those in soil in the surrounding region, scientists showed that the individual had spent his childhood more than 20 kilometres away from his place of burial.

Meanwhile, advances in instrumentation have cleared the way for fast whole genome DNA sequencing for ancient and modern samples.

The debate over the environmental impact of first peoples continues, but evidence of a human role has been mounting. In 2008, a team led by the ANU's Linda Ayliffe dated the youngest articulated megafauna remains from the key Tight Entrance Cave site in Western Australia to 48,000 to 50,000 years old.

And research led by the University of Exeter's Chris Turney suggested that a hunting blitzkrieg wiped out the Tasmanian mega-fauna. Published in PNAS, the study put the extinction to within 2,000 years of the arrival of people on the southern landmass via a temporary landbridge 43,000 years ago.

There had been a major development a year before, in 2007. In response to a question at the Geological Society of Australia's Selwyn Symposium at the University of Melbourne, Judith Field revealed long-awaited direct uranium-series–ESR combo results on megafauna teeth from Cuddie Springs. The dates, as predicted by Gillespie and Brook, were 10,000 years older than radiocarbon and OSL dates from the archaeological excavations, pouring cold water on the argument for a long overlap between megafauna and people.

However, at the symposium, Stephen Wroe was still alleging political motives. 'What concerns me is what I think is essentially a political drive to establish one or the other really exposes us to the risk of blinding ourselves to all the contingencies,' he said. 'In a world where, right now, climate change is arguably the biggest single problem facing mankind, arguments and lines of investigation that seek to sideline all that are potentially dangerous.'

'You're treading on dangerous ground here,' Gifford Miller retorted. 'You're implying that the [scientific] community has got an agenda.'

Epilogue

Set in the earth to insulate its precious contents from the temperature extremes outside, the building on Aboriginal land overlooking the lunette would reflect the 'new agenda' since Mungo Lady made her appearance 40 years before.

Hopes were high that the planned keeping place would revitalise research after decades of uncertainty. Scientists could perhaps go there to study Aboriginal remains and artefacts repatriated to the Willandra Lakes from institutions around Australia and the world. And the Three Traditional Tribal Groups would have the remains back in country. Little research had been done on human remains from Lake Mungo—or other sites—over the preceding 20 years. The politics surrounding human burials got too hot. Although the keeping place would not guarantee research—that would be in the hands of the 3TTGs—it would not block it.

The $4 million facility, which was planned to be commissioned in 2008, would have state-of-the-art environmental controls for the collection, along with a laboratory and education centre. It would generate jobs for the locals, and in 2006 the National Museum of Australia was already training young Aboriginal Australians in conservation. The 3TTGs had chosen Greg Burgess to design the building. The architect was well known for his design of the Uluru–Kata Tjuta cultural centre, a structure embedded in what his company has described as a 'delicate environment, both ecologically and politically'. His conceptual design of a circular structure for Mungo was received warmly at the 2006 Legacy of an Ice Age conference by veteran researchers hoping to hand over to younger colleagues. The keeping place seemed solid insurance against the loss of their intellectual legacy.

Opinions among the 3TTGs on research on human remains has varied. Asked if she would support further research on Mungo Lady and Mungo Man, Junette Mitchell said: 'They've done enough research on them.'

Meanwhile, the room at the ANU that houses the Willandra Lakes skeletons, over which Alan Thorne, the Paakantji, Mutthi Mutthi and Ngyiampaa have joint curatorship, is rarely unlocked these days.

Erosion of the dunes constantly exposes more skeletons in the Willandra Lakes, and heritage managers, who attempt to preserve the remains in the ground, know of many. Salvage excavations of disturbed burials saved a few skeletons from the elements, and they are kept at Mungo.

At the time of writing, Mungo Child remained in the sand, and scientists feared the skeleton would be too far gone to give much away about childhood in Ice Age Willandra.

In 2007, the Australian Government knocked back an application for funding for the Willandra Lakes keeping place.

***

Forty thousand years ago, the death of a young woman sparked a lakeside ritual that lasted for days. Her short life had centred on five vast inland lakes fringed with brilliant gold wattle. She had watched the sun rise over huge, pale, crescent-shaped sand dunes on the eastern shores of Lake Mungo. She had seen the sky set ablaze in fantastic but portentous sunsets as the wind blew fine red desert dust from the Centre.

Mungo Lady was again at the centre of a ceremony in 1990, when her remains were repatriated to the 3TTGs, and she now lies in a temporary keeping place in Mungo Visitors' Centre.

Junette Mitchell is still waiting for an answer to the question about her relationship to Mungo Lady. She still tells the story of how, on the day of the Ice Age lady's return to country, she saw a willy-willy whip up sand over the Mungo lunette.

Notes

1 Timelords and god-scientists
1. J. Mulvaney, pers. comm., September 2008.

2 Heat and light
1. J. Mulvaney & J. Kamminga, *Prehistory of Australia*, Allen & Unwin, Sydney, 1999, p. 2.

5 The melée
1. P. Vickers-Rich & T. Rich, *Wildlife of Gondwana*, Reed, Chatswood, 1993, p. 54.
2. D. Merrilees, pers. comm., May 2005.
3. E. Lundelius, pers. comm., June 2005.
4. Science Show, ABC Radio National, September 6, 2001, <www.abc.net.au/rn/science/ss/stories/s356397.htm>

6 Inside Geny's eggshell
1. Based on: P.F. Murray & P. Vickers-Rich, *Magnificent Mihirungs*, Indiana University Press, Bloomington and Indianapolis, USA, 2004; T. Flannery, *The Future Eaters*, Reed Books, Chatswood, 1994; R. Tedford & R. Wells, 'Pleistocene deposits and fossil vertebrates from the "Dead Heart of Australia"', *Memoirs of the Queensland Museum*, vol. 28, no. 1, pp. 263–84; Magee et al., 'Continuous 150 k.y. monsoon record from Lake Eyre, Australia: Insolation forcing implications and unexpected Holocene failure', *Geology*, vol. 32., no. 10, Oct. 2004, pp. 885–8.

7   Frank the Diprotodon

1. L. Dunn, pers. comm., Jan. 2005.
2. S. Brook, 'Battle of the mega-marsupials', *The Weekend Australian*, June 9, 2001, p. 3.
3. ibid.
4. ibid.

8   Silicon beasts

1. G. Prideaux, pers. comm., Jan. 2007.
2. R. Roberts, pers. comm., Jan. 2007.

13   Cool science, hot politics

1. T. Flannery, pers. comm., April 2, 2005.
2. P. Martin, pers. comm., June 2005.

15   Gene wars

1. C. Jones and J. Dixon, 'Who was Mungo's Mum?' *Canberra Times*, Jan. 11, 2001, Science page.
2. C. Jones, 'Challenge for Mungo skeleton theory', *Australian Financial Review*, Nov. 1, 2001, p. 58.

16   Roots

1. R. Clarke, pers. comm., Nov. 3, 2005.
2. T. Partridge, pers. comm., Nov. 3, 2005.
3. P.V. Tobias, pers. comm., Nov. 4, 2005.

17   Hobbit

1. <www.waspress.co.uk/journals/beforefarming/journal_20044/news/2004_4_01.pdf>
2. P. Brown, pers. comm., July 2007.

18   Neanderthal

1. C. Stringer, pers. comm., Jan. 7, 2007.
2. J. Zilhao, 'The Neanderthals: Human ancestors or aliens from outer time?', *Re: search*, University of Bristol, April 2007.
3. F. d'Errico, pers. comm., August 2007.

4. Mousterian stone tools swept across Europe in the Middle Palaeolithic, 250,000–40,000 years ago. The culture was characterised by small, precise instruments, including handaxes.

5. The Aurignacian culture of the Upper Palaeolithic, after 40,000 years ago, included sophisticated stone tools, paintings, engravings and sculptures.

6. C. Jones, 'Prehistory lessons', *Financial Times Magazine*, November 24/25, 2007, pp. 34–7.

7. ibid.

8. ibid.

9. S. Pääbo, pers. comm., October 23, 2007.

10. Sean Carroll, *Endless Forms Most Beautiful*, W.W. Norton & Co., New York, 2005.

19 'Vampire' project

1. S. Wells, pers. comm., July 2007.

2. In Australia, the most conservative political party.

3. C. Jones, 'Vampire project threatens Aboriginal beliefs on creation', *Canberra Times*, December 30, 1995, Saturday Forum, p. 10.

4. C. Jones, 'Genes study treads a tightrope', *Canberra Times*, July 19, 1997, Saturday Forum, p. 13.

5. Luigi Luca Cavalli-Sforza, pers. comm., July 2008.

6. Stone et al., 2007.

7. S. Longstaff, pers. comm., July 2, 2007.

8. S. Wells, pers. comm., July 6, 2007.

20 Back to country

1. A spokeswoman for Melbourne Museum, pers. comm., October 17, 2006.

2. C. Jones, 'Home is where the oxygen isotope is', *Australian Financial Review*, Nov. 1–2, 2003, p. 37.

3. C. Jones, 'Bones of contention', *The Bulletin*, April 9, 2002, pp. 37–9.

4. ibid.

5. N. Levitt, 'Kennewick Man: Burying the truth about America's past', spiked-science, January 20, 2001 <www.spiked-online.com/Articles/0000000053AD.htm>.

6. A spokesman for the Department of Families, Community Services and Indigenous Affairs, pers. comm., June 6, 2007.

7. C. Stringer, pers. comm., August 3, 2007.

8. N. Levitt, 'Kennewick Man: Burying the truth about America's past', spiked-science, January 20, 2001 <www.spiked-online.com/Articles/0000000053AD.htm>.

9. P.V. Tobias, pers. comm., April 2008.

10. P.V. Tobias, *Into the Past: A memoir*, Picador Africa, 2005.

11. ibid.

Bibliography

Junette

Du Cros, H., *Much More than Stones and Bones: Australian archaeology in the late twentieth century*, Melbourne University Press, 2002, p. 204.

Horton, D., *The Encyclopaedia of Aboriginal Australia*, Aboriginal Studies Press, Canberra, 1994.

I LANDFALL

1 Timelords and god-scientists

Westaway, M., 'The Pleistocene Remains Collection from the Willandra Lakes World Heritage Area, Australia, and its role in understanding modern human origins', in *Proceedings of the 7th and 8th Symposia on Collection Building and Natural History Studies in Asia and the Pacific Rim*, National Science Museum Monographs, vol. 34, 2006, pp. 127–38.

2 Heat and light

Bowdler, S., 'Some sort of dates at Malakunanja II: A reply to Roberts et al.', *Australian Archaeology*, vol. 32, 1991, pp. 50–1.

Campbell, J.B., N. Cole, E. Hatte, C. Tuniz & A. Watchman, 'Dating of rock surface accretions with Aboriginal paintings and engravings in North Queensland', *Tempus*, vol. 6, 1997, pp. 231–9.

David, B., R. Roberts, C. Tuniz, R. Jones & J. Head, 'New optical and radiocarbon dates from Ngarrabulgan Cave, a Pleistocene archaeological site in Australia: Implications for the comparability of time clocks and for the human colonisation of Australia', *Antiquity*, vol. 71, 1997, pp. 183–8.

Dayton, L. & J. Woodford, 'Australia's date with destiny', *New Scientist*, December 7, 1996, pp. 28–31.

Fullagar, R.L.K., D.M. Price & L.M. Head, 'Early human occupation of northern Australia: Archaeology and thermoluminescence dating of Jinmium rock-shelter, Northern Territory', *Antiquity*, vol. 70, Dec. 1996, pp. 751–73.

Jones, R., 'Ions and eons: Some thoughts on archaeological science and scientific archaeology', in *Archaeometry: An Australian perspective*, eds W. Ambrose & P. Duerden, Australian National University, Canberra, 1982, pp. 22–35.

Mulvaney, J. & J. Kamminga, *Prehistory of Australia*, Allen & Unwin, Sydney, 1999.

Pillans, B. & T. Naish, 'Defining the Quaternary', *Quaternary Science Reviews*, vol. 23, 2004, pp. 2271–82.

Roberts, R., M. Bird, J. Olley, R. Galbraith, E. Lawson, G. Laslett, H. Yoshida, R. Jones, R. Fullagar, G. Jacobsen & Q. Hua, 'Optical and radiocarbon dating at Jinmium rock shelter in northern Australia, *Nature*, vol. 393, May 1998, pp. 358–62.

Roberts, R., G. Walsh, A. Murray, J. Olley, R. Jones, M. Morwood, C. Tuniz, E. Lawson, M. Macphail, D. Bowdery & I. Naumann, 'Luminescence dating of rock art and past environments using mud-wasp nests in northern Australia', *Nature*, vol. 387, 1997, pp. 696–9.

Smith, L. & H. du Cros, 'Reflections on women in archaeology', in *Gendered Archaeology*, eds J. Balme & W. Beck, ANH Publications, Canberra, 1995, pp. 9–10.

Spooner, N., 'Human occupation at Jinmium, northern Australia: 116,000 years old or much less?', *Antiquity*, vol. 72, March 1998, pp. 173–8.

Tuniz, C., J.R. Bird, G.F. Herzog & D. Fink, *Accelerator Mass Spectrometry*, CRC Press, Boca Raton, USA, 1998.

3   Mungo Lady gets date

Bowler, J.M., H. Johnston, J.M. Olley, J.R. Prescott, R.G. Roberts, W. Shawcross & N.A. Spooner. 'New ages for human occupation and climatic change at Lake Mungo, Australia', *Nature*, vol. 421, 2003, pp. 837–40.

Bowler, J.M., R. Jones, H. Allen & A.G. Thorne, 'Pleistocene human remains from Australia: A living site and human cremation from Lake Mungo, western New South Wales', *World Archaeology*, vol. 2, 1970, pp. 39–60.

Field, J., S. Wroe & R. Fullagar, 'Blitzkrieg: Fact and fiction at Cuddie Springs, Australia', *Australasian Science*, July 2006, pp. 28–9.

Gillespie, R., 'Burnt and unburnt carbon: Dating charcoal and burnt bone from the Willandra Lakes, Australia', *Radiocarbon*, vol. 39, no. 3, 1997, pp. 239–50.

—— 'Dating the first Australians', *Radiocarbon*, vol. 44, no. 2, 2002, pp. 455–72.

Grün, R., M. Abeyratne, J. Head, C. Tuniz & R.E.M. Hedges. 'AMS ^{14}C analysis of teeth from archaeological sites showing anomalous ESR dating results', *Quaternary Science Reviews*, vol. 16, 1997, pp. 437–44.

Thorne, A., R. Grün, G. Mortimer, N.A. Spooner, J.J. Simpson, M. McCulloch, L. Taylor & D. Curnoe, 'Australia's oldest human remains: Age of the Lake Mungo 3 skeleton', *Journal of Human Evolution*, vol. 36, 1999, pp. 591–612.

Turney, C.S.M., M.I. Bird, L.K. Fifield, R.G. Roberts, M. Smith, C.E. Dortch, R. Grün, E. Lawson, L.K. Ayliffe, G.H. Miller, J. Dortch & R.G. Cresswell, 'Early human occupation at Devil's Lair, southwestern Australia 50,000 years ago', *Quaternary Research*, vol. 55, 2001, pp. 3–13.

4   Stairway to heaven

Chappell, J.M.A., 'Geology of coral terraces, Huon Peninsula, New Guinea: A study of Quaternary tectonic movements and sea-level changes', *Bulletin of the Geological Society of America*, vol. 85, 1974, pp. 553–70.

Chappell, J., 'Sea level changes forced ice breakouts in the Last Glacial cycle: New results from coral terraces', *Quaternary Science Reviews*, vol. 21, 2002, pp. 1229–40.

EPICA Community Members, 'Eight glacial cycles from an Antarctic ice core', *Nature*, vol. 429, 2004, pp. 623–8.

Hays, J.D., J. Imbrie & N.J. Shackleton, 'Variations in the earth's orbit: Pacemaker of the ice ages', *Science*, vol. 194, 1976, pp. 1121–32.

Jouzel J., V. Masson-Delmotte, O. Cattani, G. Dreyfus, S. Falourd, G. Hoffmann, B. Minster, J. Nouet, J.M. Barnola, J. Chappellaz, H. Fischer, J.C. Gallet, S. Johnsen, M. Leuenberger, L. Loulergue, D. Luethi, H. Oerter, F. Parrenin, G. Raisbeck, D. Raynaud, A. Schilt, J. Schwander, E. Selmo, R. Souchez, R. Spahni, B. Stauffer, J.P. Steffensen, B. Stenni, T.F. Stocker, J.L. Tison, M. Werner & E.W. Wolff, 'Orbital and millennial Antarctic climate variability over the past 800,000 years', *Science,* vol. 317, 2007, pp. 793–6.

O'Connor, S. & J. Chappell, 'Colonisation and coastal subsistence in Australia and Papua New Guinea: Different timing, different modes?', in C. Sand (ed.), *Pacific Archaeology: Assessments and prospects*, Département Archéologie, Noumea, 2003, pp. 17–32.

Raymo, M.E. & P. Huybers, 'Unlocking the mysteries of the ice ages', *Nature,* vol. 451, 2008, pp. 284–5.

Veeh, H.H. & J. Chappell, 'Astronomical theory of climatic change: Support from New Guinea, *Science*, vol. 167, 1970, pp. 862–5.

II EXTINCTION

5 The melée

Flannery, T., *The Future Eaters*, Reed Books, Chatswood, 1994.

Jones, R., 'The geographical background to the arrival of man in Australia and Tasmania', *Archaeology and Physical Anthropology in Oceania*, vol. 3, no. 3, Oct. 1968, pp. 186–215.

Long, J.A., P. Vickers-Rich, K. Hirsch, E. Bray & C. Tuniz, 'The Cervantes egg: An early Malagasy tourist to Australia', *Records of the Australian Museum*, vol. 19, 1998, pp. 39–46.

Merrilees, D., 'Man the destroyer: Late Quaternary changes in the Australian marsupial fauna', *Journal of the Royal Society of Western Australia*, vol. 51, 1968, pp. 1–24.

Roberts, R.G., T.F. Flannery, L.K. Ayliffe, H. Yoshida, J.M. Olley, G.J. Prideaux, G.M. Laslett, A. Baynes, M.A. Smith, R. Jones & B.L. Smith, 'New ages for the last Australian megafauna: Continent-wide extinction about 46,000 years ago', *Science*, vol. 292, June 2001, pp. 1888–92.

Smith, L., 'What is this thing called post-processual archaeology ... and what is its relevance to Australian archaeology?', *Australian Archaeology*, vol. 40, 1995, pp. 28–32.

6   Inside Geny's eggshell

Flannery, T., *The Future Eaters*, Reed Books, Chatswood, 1994.

Magee, J.W., G.H. Miller, N.A. Spooner & D. Questiaux, 'Continuous 150 k.y. monsoon record from Lake Eyre, Australia: Insolation-forcing implications and unexpected Holocene failure', *Geology*, vol. 32, no. 10, Oct. 2004, pp. 885–7.

Miller, G.H., J.W. Magee, B.J. Johnson, M.L. Fogel, N.A. Spooner, M.T. McCulloch & L.K. Ayliffe, 'Pleistocene extinction of *Genyornis newtoni*: Human impact on Australian megafauna', *Science*, vol. 283, Jan. 1999, pp. 205–8.

Miller, G.H., M.L. Fogel, J.W. Magee, M.K. Gagan, S.J. Clarke & B.J. Johnson, 'Ecosystem collapse in Pleistocene Australia and a human role in megafaunal extinction', *Science*, vol. 309, July 2005, pp. 287–90.

Miller, G.H., J. Mangan, D. Pollard, S. Thompson, B. Felzer & J.W. Magee, 'Sensitivity of the Australian Monsoon to insolation and vegetation: Implications for human impact on continental moisture balance', *Geology*, vol. 33, no. 1, Jan. 2005, pp. 65–8.

Murray, P.F. & P. Vickers-Rich, *Magnificent Mihirungs*, Indiana University Press, Bloomington, USA, 2004.

Tedford, R.H. & R.T. Wells, 'Pleistocene deposits and fossil vertebrates from the "dead heart" of Australia', *Memoirs of the Queensland Museum*, vol. 28, no. 1, 1990, pp. 263–84.

7 Frank the Diprotodon

Ayliffe, L.K., P.C. Marianelli, K.C. Moriarty, R.T. Wells, M.T. McCulloch, G.E. Mortimer & J. Hellstrom, '500 ka precipitation record from southeastern Australia: Evidence for interglacial relative aridity', *Geology*, vol. 26, no. 2, February 1998, pp. 147–50.

Brook, B.W., D.M.J.S. Bowman, D.A. Burney, T.F. Flannery, M.K. Gagan, R. Gillespie, C.N. Johnson, P. Kershaw, J.W. Magee, P.S. Martin, G.H. Miller, B. Peiser & R.G. Roberts, 'Would the Australian megafauna have become extinct if humans had never colonised the continent? Comments on "A review of the evidence for a human role in the extinction of Australian megafauna and an alternative explanation" by S. Wroe and J. Field', *Quaternary Science Reviews*, vol. 26, no. 3–4, 2007, pp. 560–4.

Dodson, J., R. Fullagar, J. Furby, R. Jones & I. Prosser, 'Humans and megafauna in a Late Pleistocene environment at Cuddie Springs, north-western New South Wales', *Archaeology in Oceania*, vol. 28, 1993, pp. 93–9.

Field, J., R. Fullagar & G. Lord, 'A large area archaeological excavation at Cuddie Springs', *Antiquity*, vol. 75, no. 290, Dec. 2001, pp. 696–702.

Gillespie, R. & B.W. Brook, 'Is there a Pleistocene archaeological site at Cuddie Springs?', *Archaeology in Oceania*, vol. 41, 2006, pp. 1–11.

Gillespie, R., B.W. Brook & A. Baynes, 'Short overlap of humans and megafauna in Pleistocene Australia', *Alcheringa* Special Issue, vol. 1, 2006, pp. 163–86.

Grün, R., R. Wells, S. Eggins, N. Spooner, M. Aubert, L. Brown & E. Rhodes, 'Electron spin resonance dating of South Australian megafauna sites', *Australian Journal of Earth Sciences*, vol. 55, no. 6/7, 2008, pp. 917–35.

Horton, D., *The Pure State of Nature*, Allen & Unwin, Sydney, 2000, p. 115.

O'Connor, S. & J. Chappell, 'Colonisation and coastal subsistence in Australia and Papua New Guinea: Different timing, different modes?', in C. Sand (ed.), *Pacific Archaeology: Assessments and prospects*, Départment Archéologie, Noumea, 2003, pp. 17–32.

Pate, F.D., M.C. McDowell, R.T. Wells & A.M. Smith, 'Last recorded evidence for megafauna at Wet Cave, Naracoorte, South Australia 45,000 years ago', *Australian Archaeology*, no. 54, 2002, pp. 53–5.

Prideaux G.J., R.G. Roberts, D. Megirian, K.E. Westaway, J.C. Hellstrom & J.M. Olley, 'Mammalian responses to Pleistocene climate change in southeastern Australia', *Geology*, vol. 35, no. 1, 2007, pp. 33–6.

Roberts, R.G., T.F. Flannery, L.K. Ayliffe, H. Yoshida, J.M. Olley, G.J. Prideaux, G.M. Laslett, A. Baynes, M.A. Smith, R. Jones & B.L. Smith, 'New ages for the last Australian megafauna: Continent-wide extinction about 46,000 years ago', *Science*, vol. 292, June 2001, pp. 1888–92.

Trueman, C.N.G., J.H. Field, J. Dortch, B. Charles & S. Wroe, 'Prolonged coexistence of humans and megafauna in Pleistocene Australia', *PNAS*, vol. 102, no. 23, 2005, pp. 8381–5.

van Huet, S., R. Grün, C.V. Murray-Wallace, N. Redvers-Newton & J.P. White, 'Age of Lancefield megafauna: A reappraisal', *Australian Archaeology*, no. 46, 1998, pp. 5–11.

Wells, R.T., R. Grün, J. Sullivan, M. Forbes, S. Dalgairns, E. Bestland, N. Spooner, K.E. Walshe, E. Rhodes & S. Eggins, 'Late Pleistocene megafauna site at Black Creek Swamp, Flinders Chase National Park, Kangaroo Island, South Australia', *Alcheringa*, Special Issue, vol. 1, 2006, pp. 367–87.

Wroe, S. & J. Field, 'A reply to comment by Brook et al. "Would the Australian megafauna have become extinct if humans had never colonised the continent?' *Quaternary Science Reviews*, vol. 26, 2007, pp. 565–7.

8   Silicon beasts

Alroy, J., 'A multispecies overkill simulation of the end-Pleistocene megafaunal mass extinction', *Science*, vol. 292, June 2001, pp. 1893–96.

Barnosky, A., P.L. Koch, R.S. Feranec, S.L. Wing & A.B. Shabel, 'Assessing the causes of Late Pleistocene extinctions on the continents', *Science*, vol. 306, no. 5693, 2004, pp. 70–5.

Brook, B.W. & D.M.J.S. Bowman, 'The uncertain blitzkrieg of

Pleistocene megafauna', *Journal of Biogeography*, vol. 31, 2004, pp. 517–23.

Brook, B.W. & C.N. Johnson, 'Selective hunting of juveniles as a cause of the imperceptible overkill of the Australian Pleistocene megafauna', *Alcheringa*, Special Issue, vol. 1, 2006, pp. 39–48.

Kershaw, A.P., 'Climatic change and Aboriginal burning in north-east Australia during the last two glacial/interglacial cycles', *Nature*, vol. 322, July 1986, pp. 47–9.

Prideaux, G.J., J. A. Long, L.K. Ayliffe, J.C. Hellstrom, B. Pillans, W.E. Boles, M.N. Hutchinson, R.G. Roberts, M.L. Cupper, L.J. Arnold, P.D. Devine & N.M. Warburton, 'An arid-adapted middle Pleistocene vertebrate fauna from south-central Australia', *Nature*, vol. 445, January 2007, pp. 422–5.

Tuck, G.N., T. Polacheck, J.P. Croxall & H. Weimerskirch, 'Modelling the impact of fishery by-catches on albatross populations', *Journal of Applied Ecology*, vol. 38, 2001, pp. 1182–96.

Turney, C.S.M., A.P. Kershaw, P. Moss, M.I. Bird, L.K. Fifield, R.G. Cresswell, G.M. Santos, M.L. Di Tada, P. A. Hausladen & Y. Zhou, 'Redating the onset of burning at Lynch's Crater (North Queensland): Implications for human settlement in Australia', *Journal of Quaternary Science*, vol. 16, no. 8, 2001, pp. 767–71.

Turney, C.S.M., A.P. Kershaw, S.C. Clemens, N. Branch, P.T. Moss & L.K. Fifield, 'Millennial and orbital variations of El Niño/ Southern Oscillation and high-latitude climate in the last glacial period', *Nature*, vol. 428, March 2004, pp. 306–10.

9   New World order

Bahn P., 'Dating the first American: When did people first enter the New World? Clues from distinctive rock art may push the date back by thousands of years', *New Scientist*, vol. 1778, July 1991, p. 26.

Barnosky, A., P.L. Koch, R.S. Feranec, S.L. Wing & A.B. Shabel, 'Assessing the causes of Late Pleistocene extinctions on the continents', *Science*, vol. 306, no. 5693, 2004, pp. 70–5.

Gilbert, M.T.P., L.P. Tomsho, S. Rendulic, M. Packard, D.I. Drautz, A. Sher, A. Tikhonov, L. Dalén, T. Kuznetsova, P. Kosintsev,

P.F. Campos, T. Higham, M.J. Collins, A.S. Wilson, F. Shidlovskiy, B. Buigues, P.G.P. Ericson, M. Germonpré, A. Götherström, P. Iacumin, V. Nikolaev, M. Nowak-Kemp, E. Willerslev, J.R. Knight, G.P. Irzyk, C.S. Perbost, K.M. Fredrikson, T.T. Harkins, S. Sheridan, W. Miller & S.C. Schuster, 'Whole-genome shotgun sequencing of mitochondria from ancient hair shafts', *Science*, vol. 317, no. 5846, 2007, pp. 1927–30.

Gilbert, M.T.P., et al., 'Palaeo-Eskimo mtDNA genome reveals matrilineal discontinuity in Greenland', *Sciencexpress*, May 29, 2008, pp. 1–4.

Goebel, T., M.R. Waters & D.H. O'Rourke, 'The late Pleistocene dispersal of modern humans in the Americas', *Science*, vol. 319, 2008, pp. 1497–1502.

Hajdas, I., D.J. Lowe, R.M. Newnham & G. Bonani, 'Timing of the Late-Glacial climate reversal in the Southern Hemisphere using high-resolution radiocarbon chronology for Kaipo bog, New Zealand', *Quaternary Research*, vol. 65, no. 2, 2006, pp. 340–5.

Martin, P.S., 'The discovery of America', *Science*, vol. 179, no. 4077, March 1973, pp. 969–74.

Martin, P.S., *Twilight of the Mammoths: Ice Age extinctions and the rewilding of America*, Berkeley, University of California Press, 2005.

Waters, M.R. & T.W. Stafford Jr., 'Redefining the age of Clovis: Implications for the peopling of the Americas', *Science*, vol. 315, 2007, pp. 1122–6.

Webb, S., *The First Boat People*, Cambridge University Press, Cambridge, 2007, p. 318.

10 Blast from the past

Pionar, H., M. Kuch, G. McDonald, P. Martin & S. Pääbo, 'Nuclear gene sequences from a late Pleistocene sloth coprolite', *Current Biology*, vol. 13, 2003, pp. 1150–2.

Steadman, D.W., P.S. Martin, R.D.E. MacPhee, A.J.T. Jull, H.G. McDonald, C.A. Woods, M. Iturralde-Vinent & G.W.L. Hodgins, 'Asynchronous extinction of Late Quaternary sloths on continents and islands', *PNAS*, vol. 102, no. 33, 2005, pp. 11,763–8.

11 Bison

Drummond, A.J., A. Rambaut, B. Shapiro & O.G. Pybus, 'Bayesian coalescent inference of past population dynamics from molecular sequences', *Molecular Biology and Evolution*, vol. 22, no. 5, 2005, pp. 1185–92.

Shapiro B., A.J. Drummond, A. Rambaut, M.C. Wilson, P.E. Matheus, A.V. Sher, O.G. Pybus, M.T.P. Gilbert, I. Barnes, J. Binladen, E. Willerslev, A.J. Hansen, G.F. Baryshnikov, J.A. Burns, S. Davydov, J.C. Driver, D.G. Froese, C.R. Harington, G. Keddie, P. Kosintsev, M.L. Kunz, L.D. Martin, R.O. Stephenson, J. Storer, R. Tedford, S. Zimov & A. Cooper, 'Rise and fall of the Beringian steppe Bison', *Science*, vol. 306, 2004, pp. 1561–5.

Willerslev, E., A. Hansen, J. Binladen, T.B. Brand, M.T.P. Gilbert, B. Shapiro, M. Bunce, C. Wiuf, D.A. Gilichinsky & A. Cooper, 'Diverse plant and animal genetic records from Holocene and Pleistocene sediments', *Science*, vol. 300, no. 5620, 2003, pp. 791–5.

12 Cosmic impact

Chyba, C.F., P.J. Thomas & K.J. Zahnle, 'The 1908 Tunguska explosion: Atmospheric disruption of a stony asteroid', *Nature*, vol. 361, 1993, pp. 40–4.

Firestone, R.B., A. West, J.P. Kennett, L. Becker, T.E. Bunch, Z.S. Revay, P.H. Schultz, T. Belgya, D.J. Kennett, J.M. Erlandson, O.J. Dickenson, A.C. Goodyear, R.S. Harris, G.A. Howard, J.B. Kloosterman, P. Lechler, P.A. Mayewski, J. Montgomery, R. Poreda, T. Darrah, S.S. Que Hee, A.R. Smith, A. Stich, W. Topping, J.H. Wittke & W.S. Wollbach, 'Evidence for an extraterrestrial impact 12,900 years ago that contributed to the megafaunal extinctions and the Younger Dryas cooling', PNAS, vol. 104, no. 41, 2007, pp. 16,016–21.

Kerr, R.A., 'Experts find no evidence for a mammoth-killer impact', *Science*, vol. 319, 2008, pp. 1331–2.

Longo, G., R. Serra, S. Cecchini & M. Galli, 'Search for microremnants of the Tunguska cosmic body', *Planetary and Space Science*, vol. 42, no. 2, 1994, pp. 163–77.

13 Cool science, hot politics

Benson, J., 'Beautiful lies', *Quarterly Essay*, vol. 13, 2004, p. 127.

Brook, B.W., D.M.J.S. Bowman, D.A. Burney, T.F. Flannery, M.K. Gagan, R. Gillespie, C.N. Johnson, P. Kershaw, J.W. Magee, P.S. Martin, G.H. Miller, B. Peiser & R.G. Roberts, 'Would the Australian megafauna have become extinct if humans had never colonised the continent? Comments on "A review of the evidence for a human role in the extinction of Australian megafauna and an alternative explanation" by S. Wroe and J. Field', *Quaternary Science Reviews*, vol. 26, no. 3–4, 2007, pp. 560–4.

Fiedel, S. & G. Haynes, 'A premature burial: Comments on Grayson and Meltzer's requiem for overkill', *Journal of Archaeological Science*, vol. 31, 2004, pp. 121–31.

Flannery, T., 'Beautiful lies: Response to correspondence', *Quarterly Essay*, vol. 13, 2004, p. 135.

Grayson, D. & D. Meltzer, 'A requiem for North American overkill', *Journal of Archaeological Science*, vol. 30, 2003, pp. 585–93.

Head, L., 'Meganesian barbeque', *Meanjin*, vol. 54, 1995, pp. 702–9.

Salleh, A., 'Early fire-farmers switched off monsoon', ABC, 2005 <www.abc.net.au/science/news/stories/s1289939.htm>.

Sheehan, P., 'The crucifixion of St Timothy', *The Sydney Morning Herald*, June 5–6, 2004, Spectrum, p. 4.

Smith, L. & H. du Cros, 'Reflections on women in archaeology', in *Gendered Archaeology*, eds J. Balme & W. Beck, ANH Publishing, Canberra, 1995, p. 10.

Surovell, T.A. & N.M. Waguespack, 'How many elephant kills are 14? Clovis mammoth and mastodon kills in context', *Quaternary International*, vol. 191, 2008, pp. 82–97.

Wroe, S., J. Field, R. Fullagar & L.S. Jermiin, 'Megafaunal extinction in the late Quaternary and the global overkill hypothesis', *Alcheringa*, vol. 28, no. 1, 2004, pp. 291–331.

Wroe, S., 'On little lizards and the big extinction blame game', *Quaternary Australasia*, vol. 23, no. 1, July 2005, pp. 8–11.

Wroe, S., J. Field & R. Fullagar, 'Lost giants', *Nature Australia*, vol. 27, no. 5, 2002, pp. 54–61.

III ORIGINS

15 Gene wars

Adcock, G.J., E.S. Dennis, S. Easteal, G.A. Huttley, L.S. Jermiin, W.J. Peacock & A.G. Thorne, 'Mitochondrial DNA sequences in ancient Australians: Implications for modern human origins', PNAS, vol. 98, no. 2, 2001, pp. 537–42.

Brown, P. 'The First Australians', *Australasian Science*, vol. 21, May 2000, pp. 28–31.

Cameron, D.W. & C.P. Groves, *Bones, Stones and Molecules: 'Out of Africa' and human origins*, Elsevier Academic Press, Amsterdam, 2004.

Cann, R.L., M. Stoneking & A.C. Wilson, 'Mitochondrial DNA and human evolution', *Nature*, vol. 325, January 1, 1987, pp. 31–6.

Cooper, A., A. Rambaut, V. Macaulay, E. Willersley, A.J. Hansen, C. Stringer, 'Human origins and ancient human DNA', *Science*, vol. 292, issue 5522, June 2001, pp. 1655–6.

Underhill, P.A., Peidong Shen, A.A. Lin, Li Jin, G. Passarino, Wei H. Yang, E. Kauffman, Batsheva Bonné-Tamir, J. Bertranpetit, P. Francalacci, Muntaser Ibrahim, Trefor Jenkins, Judith R. Kidd, S. Qasim Mehdi, M.T. Seielstad, R. Spencer Wells, A. Piazza, R.W. Davis, M.W. Feldman, L.L. Cavalli-Sforza & P.J. Oefner, 'Y chromosome sequence variation and the history of human populations', *Nature Genetics*, vol. 26, Nov. 1, 2000, pp. 358–61.

16 Roots

Alemseged, Z., F. Spoor, W. H. Kimbel, R. Bobe, D. Geraads, D. Reed & J.G. Wynn, 'A juvenile early hominin skeleton from Dikika, Ethiopia', *Nature*, vol. 443, 2006, pp. 296–301.

Anikovich, M.V., A.A. Sinitsyn, J.F. Hoffecker, V.T. Holliday, V.V. Popov, S.N. Lisitsyn, S.L. Forman, G.M. Levkovskaya, G.A. Pospelova, I.E. Kuz'mina, N.D. Burova, P. Goldberg, R.I. Macphail, B. Giaccio & N.D. Praslov, 'Early Upper Palaeolithic in Eastern Europe and implications for the dispersal of modern humans', *Science*, vol. 315, 2007, pp. 223–6.

Arzarello, M., F. Marcolini, G. Pavia, M. Pavia, C. Petronio, M. Petrucci, L. Rook & R. Sardella, 'Evidence of earliest human occurrence in Europe: The site of Pirro Nord (Southern Italy)', *Naturwissenschaften*, vol. 94, 2007, pp. 107–12.

Backwell, L.R. & F. d'Errico, 'Evidence of termite foraging by Swartkrans early hominids', PNAS, vol. 98, 2001, pp. 1358–63.

Brain, C.K., *The Hunters or the Hunted? An Introduction to African Cave Taphonomy*, University of Chicago Press, Chicago, IL, 1981.

Brunet, M., A. Beauvilain, Y. Coppens, E. Heintz, H.E. Aladji & D. Pilbeam, 'The first Australopithecine 2,500 kilometres west of the Rift Valley (Chad)', *Nature*, vol. 378, 1995, pp. 273–5.

Brunet, M., F. Guy, D. Pilbeam, H.T. Mackaye, A. Likius, D. Ahounta, A. Beauvilain, C. Blondel, H. Bocherens, J-R. Bois-serie, L. De Bonis, Y. Coppens, J. Dejax, C. Denys, P. Duringer, V. Eisenmann, G. Fanone, P. Fronty, D. Geraads, T. Lehmann, F. Lihoreau, A. Louchart, A. Mahamat, G. Merceron, G. Mouche-lin, O. Otero, P.P. Campomanes, M.P. De Leon, J-C. Rage, M. Sapanet, M. Schuster, J. Sudre, P. Tassy, X. Valentin, P. Vignaud, L. Viriot, A. Zazzo & C. Zollikofer, 'A new hominid from the Upper Miocene of Chad, Central Africa', *Nature*, vol. 418, 2002, pp. 145–51.

Cavalli-Sforza, L.L. & A.W.F. Edwards, 'Analysis of human evolu-tion', in *Genetics Today*, ed. S.J. Geerts, Pergamon Press, Oxford, vol. 2, 1964, pp. 923–33.

Clarke, R.J., 'First ever discovery of a well-preserved skull and asso-ciated skeleton of *Australopithecus*', *South African Journal of Science*, vol. 94, 1998, pp. 460–3.

—— 'Discovery of complete arm and hand of the 3.3 million-year-old *Australopithecus* skeleton from Sterkfontein, South Africa', *South African Journal of Science*, vol. 95, 1999, pp. 477–80.

Clarke, R.J. & K. Kuman, 'The Sterkfontein Caves: Palaeontological and archaeological site', <www.africangamesafari.com/kolebka.pdf>

Dart, R.A., '*Australopithecus africanus*: The man-ape of South Africa', *Nature*, vol. 115, 1925, pp. 195–9.

—— 'The predatory implemental technique of *Australopithecus*', *American Journal of Physical Anthropology*, vol. 7, 1949, pp. 1–38.

Dennell, R. & W. Roebroeks, 'An Asian perspective on early human dispersal from Africa', *Nature*, vol. 438, 2005, pp. 1099–104.

Grün, R. & A. Thorne 'Dating the Ngandong humans', *Science*, vol. 276, June, 1997, pp. 1575–6.

Holliday, V.T., J.F. Hoffecker, P. Goldberg, R.I. Macphail, S.L. Forman, M. Anikovich & A. Sinitsyn, 'Geoarchaeology of the Kostenki–Borshchevo sites, Don River Valley', *Geoarchaeology*, vol. 22, 2007, pp. 181–228.

Hudjashov, G., T. Kivisild, P.A. Underhill, P. Endicott, J.J. Sanchez, A.A. Lin, P. Shen, P. Oefner, C. Renfrew, R. Villems & P. Forster, 'Revealing the prehistoric settlement of Australia by Y chromosome and mtDNA analysis', *PNAS*, vol. 104, no. 21, May 22, 2007, pp. 8726–30.

Macauley, V., C. Hill, A. Achilli, C. Rengo, D. Clarke, W. Meehan, J. Blackburn, O. Semino, R. Scozzari, F. Cruciani, A. Taha, N. Kassim Shaari, J. Maripa Raja, P. Ismail, Z. Zainuddin, W. Goodwin, D. Bulbeck, H.-J. Bandelt, S. Oppenheimer, A. Torroni & M. Richards, 'Single, rapid coastal settlement of Asia revealed by analysis of complete mitochondrial genomes', *Science*, vol. 308, no. 5724, 2005, pp. 1034–6.

Mellars, P.A., 'A new radiocarbon revolution and the dispersal of modern humans in Eurasia', *Nature*, vol. 439, 2006, pp. 931–5.

Partridge, T.C., 'Re-appraisal of lithostratigraphy of Sterkfontein hominid site', *Nature*, vol. 275, 1978, pp. 282–7.

Partridge, T.C., J. Shaw, D. Heslop & R.J. Clarke, 'The new hominid skeleton from Sterkfontein, South Africa; age and preliminary assessment', *Journal of Quaternary Science*, vol. 14, no. 4, 1999, pp. 293–8.

Partridge, T.C., D.E. Granger, M.W. Caffee, & R.J. Clarke, 'Lower Pliocene hominid remains from Sterkfontein', *Science*, vol. 300, 2003, pp. 607–12.

Petraglia, M., R. Korisettar, N. Boivin, C. Clarkson, P. Ditchfield, S. Jones, J. Koshy, M. Mirazón Lahr, C. Oppenheimer, D. Pyle, R. Roberts, J.-L. Schwenninger, L. Arnold & K. White, 'Middle Palaeolithic assemblages from the Indian Subcontinent before and after the Toba super-eruption, *Science*, vol. 317, no. 5834, 2007, pp. 114–16.

Scholz, C.A., T.C. Johnson, A.S. Cohen, J.W. King, J.A. Peck, J.T. Overpeck, M.R. Talbot, E.T. Brown, L. Kalindekafe, P.Y.O. Amoako, R.P. Lyons, T.M. Shanahan, I.S. Castañeda, C.W. Heil, S.L. Forman, L.R. McHargue, K.R. Beuning, J. Gomez & J. Pierson, 'East African megadroughts between 135 and 75 thousand years ago and bearing on early-modern human origins', PNAS, vol. 104, 2007, pp. 16,416–21.

Spoor, F., G. Leakey, P.N. Gathogo, F.H. Brown, S.C. Antón, I. McDougall, C. Kiarie, F.K. Manthi & L.N. Leakey, 'Implications of new early *Homo* fossils from Ileret, east of Lake Turkana, Kenya', *Nature*, vol. 448, 9 Aug. 2007, pp. 688–91.

Swisher, C.G., G.H. Curtis, T. Jacob, A.G. Getty, A. Suprijo & Widiasmoro, 'Age of the earliest known hominids in Java, Indonesia', *Science*, vol. 263, 1994, pp. 1118–21.

Swisher, C.G., W.J. Rink, S.C. Antón, H.P. Schwarcz, G.H. Curtis, A. Suprijo & Widiasmoro, 'Latest *Homo erectus* at Java: Potential contemporaneity with *Homo sapiens* in Southeast Asia', *Science*, vol. 274, 1996, pp. 1870–74.

Tobias, P.V., *Into the Past: A Memoir*, Wits University Press, Johannesburg, 2005.

Turney, C.S.M., R.G. Roberts & Z. Jacobs, 'Archaeology: Progress and pitfalls in radiocarbon dating', *Nature*, vol. 443, no. 7108, 2006, p. E3.

Vaks, A., M. Bar-Matthews, A. Ayalon, A. Matthews, L. Halicz & A. Frumkin, 'Desert speleothems reveal climatic windows for African exodus of early modern humans', *Geology*, vol. 35, 2007, pp. 831–4.

Vekua, A., D. Lordkipanidze, G.P. Rightmire, J. Agusti, R. Ferring, G. Maisuradze, A. Mouskhelishvili, M. Nioradze, M. Ponce de Leon, M. Tappen, M. Tvalchrelidze & C. Zollikofer, 'A new skull of early *Homo* from Dmanisi, Georgia', *Science*, vol. 97, 2002, pp. 85–9.

Webb, S., *The First Boat People*, Cambridge University Press, Cambridge, 2007, p. 318.

Wood, B. & M. Collard, 'The human genus', *Science*, vol. 284, 1999, pp. 65–71.

Wynn, J.G., Z. Alemseged, R. Bobe, D. Geraads, D. Reed & D.C. Roman, 'Geological and palaeontological context of a Pliocene juvenile hominin at Dikika, Ethiopia', *Nature*, vol. 443, no. 7109, 2006, pp. 332–6.

Zhivotovsky, L.A., N.A. Rosenberg & M.W. Feldman, 'Features of evolution and expansion of modern humans, inferred from genome-wide microsatellite markers', *American Journal of Human Genetics*, vol. 72, 2003, pp. 1171–86.

17 Hobbit

Berger, L.R., S.E. Churchill, B. De Klerk & R.L. Quinn, 'Small-bodied humans from Palau, Micronesia', *PLOS ONE*, 2008, 3(3): e1780. doi:10.1371/journal.pone.0001780.

Brown, P., T. Sutikna, M.J. Morwood, R.P. Soejono, E. Jatmiko, Wayhu Saptomo & Rokus Awe Due, 'A new small-bodied hominin from the Late Pleistocene of Flores, Indonesia', *Nature*, vol. 431, 2004, pp. 1055–61.

Cooper, A., A. Rambaut, V. Macaulay, E. Willerslev, J. Hansen & C. Stringer, 'Human origins and ancient human DNA', *Science*, vol. 292, no. 5522, 2001, pp. 1655–6.

Culotta, E., 'The fellowship of the Hobbit', *Science*, vol. 317, 2007, pp. 740–2.

Dalton, R., 'Bones, isles and videotape', *Nature*, vol. 452, April 17, 2008, pp. 806–8.

Dennell, R.W. & W. Roebroeks, 'An Asian perspective on early human dispersal from Africa', *Nature*, vol. 438, 2005, pp. 1099–104.

Falk, D., C. Hildebolt, K. Smith, M.J. Morwood, T. Sutikna, P. Brown, Jatmiko, E. Wayhu Saptomo, B. Brunsden & F. Prior, 'The brain of LB1, *Homo floresiensis*', *Science*, vol. 308, April 2005, pp. 242–5.

Falk D., C. Hildebolt, K. Smith, M.J. Morwood, T. Sutikna, Jatmiko, E. Wayhu Saptomo, H. Imhof, H. Seidler & F. Prior, 'Brain shape in human microcephalics and *Homo floresiensis*', PNAS, vol. 104, 2007, pp. 2513–18.

Henneberg, M. & J. Schofield, *The Hobbit Trap: Money, fame, science and the discovery of a 'new species'*, Wakefield Press, South Australia, 2008.

Hershkovitz, I., L. Kornreich & Z. Laron, 'Comparative skeletal features between *Homo floresiensis* and patients with primary growth hormone insensitivity (Laron Syndrome)', *American Journal of Physical Anthropology*, vol. 134, no. 2, 2007, pp. 198–208.

Jacob, T., E. Indriati, R.P. Soejono, K. Hsü, D.W. Frayer, R.B. Eckhardt, A.J. Kuperavage, A. Thorne & M. Henneberg, 'Pygmoid Australomelanesian *Homo sapiens* skeletal remains from Liang Bua, Flores: Population affinities and pathological abnormalities', PNAS, vol. 103, 2006, pp. 13421–6.

Larson, S.G., W.L. Jungers, M.J. Morwood, T. Sutikna, Jatmiko, E.W. Saptomo, R.A. Due & T. Djubiantono, '*Homo floresiensis* and the evolution of the hominin shoulder', *Journal of Human Evolution*, vol. 53, 2007, pp. 718–31.

Morwood, M.J., P.B. O'Sullivan, F. Aziz & A. Raza, 'Fission-track ages of stone tools and fossils on the east Indonesian island of Flores', *Nature*, vol. 392, 1998, pp. 173–6.

Morwood, M.J., R.P. Soejono, R.G. Roberts, T. Sutikna, C.S.M. Turney, K.E. Westaway, W.J. Rink, J.-X. Zhao, G.D. van den Bergh, Rokus Awe Due, D.R. Hobbs, M.W. Moore, M.I. Bird & L.K. Fifield, 'Archaeology and age of a new hominin from Flores in eastern Indonesia', *Nature*, vol. 431, 2004, pp. 1087–91.

Morwood, M. & P. Van Oosterzee, *The Discovery of the Hobbit*, Random House, Sydney, 2007.

Obendorf, P.J., C.E. Oxnard & B.J. Kefford, 'Are the small human-like fossils found on Flores human endemic cretins?', *Proceedings of the Royal Society of London*, B. 275, 2008, pp. 1287–96.

Tocheri, M.W., C.M. Orr, S.G. Larson, T. Sutikna, Jatmiko, E. Wahyu Saptomo, R. Awe Due, T. Djubiantono, M.J. Morwood & W.L. Jungers, 'The primitive wrist of *Homo floresiensis* and its implications for hominin evolution', *Science*, vol. 317. no. 5845, September 21, 2007, pp. 1743–5.

van Heteren, A.H., '*Homo floresiensis* as an island form', *Palarch's Journal of Vertebrate Palaeontology*, vol. 5, 2008, pp. 1–19.

Wheeler, D.A., M. Srinivasan, M. Egholm, Y. Shen, L. Chen, A. McGuire, W. He, Y. Chen, V. Makhijani, G.T. Roth, X. Gomes,

K. Tartaro, F. Niazi, C.L. Turcotte, G.P. Irzyk, J.R. Lupski, C. Chinault, X. Song, Y. Liu, Y. Yuan, L. Nazareth, X. Qin, D.M. Muzny, M. Margulies, G.M. Weinstock, R.A. Gibbs & J.M. Rothberg, 'The complete genome of an individual by massively parallel DNA sequencing', *Nature*, vol. 452, April. 2008, pp. 872–6.

Zhu R.X., R. Potts, F. Xie, K.A. Hoffman, C.L. Deng, C.D. Shi, Y.X. Pan, H.Q. Wang, R.P. Shi, Y.C. Wang, G.H. Shi & N.Q. Wu, 'New evidence on the earliest human presence at high northern latitudes in northeast Asia', *Nature*, vol. 431, 2004, pp. 559–62.

18 Neanderthal

Carbonell, E., J.M. Bermúdez de Castro, J.M. Parés, A. Pérez-González, G. Cuenca-Bescós, A. Ollé, M. Mosquera, R. Huguet, J. van der Made, A. Rosas, R. Sala, J. Vallverdú, N. García, D.E. Granger, M. Martinón-Torres, X.P. Rodríguez, G.M. Stock, J.M. Vergès, E. Allué, F. Burjachs, I. Cáceres, A. Canals, A. Benito, C. Díez, M. Lozano, A. Mateos, M. Navazo, J. Rodríguez, J. Rosell & J.L. Arsuaga, 'The first hominin of Europe', *Nature*, vol. 452, no. 7186, 2008, pp. 465–9.

Huxley, T.H., *Evidence as to Man's Place in Nature*, D. Appleton and Company, New York, 1863.

Krings, M., A. Stone, R.W. Schmitz, H. Krainitzki, M. Stoneking & S. Pääbo, 'Neanderthal DNA sequences and the origins of modern humans', *Cell*, vol. 90, 1997, pp. 19–30.

Lewin, R. & R.A. Foley, *Principles of Human Evolution*, Blackwell, Oxford, 2004.

Martinon-Torres, M., M.J. Bermúdez de Castro, A. Gómez-Robles, J.L. Arsuaga, E. Carbonell, D. Lordkipanidze, G. Manzi & A. Margvelashvili, 'Dental evidence on the hominin dispersals during the Pleistocene', PNAS, vol. 104, no. 33, 2007, pp. 13,279–82.

Moser, S., 'The visual language of archaeology: A case study of the Neanderthals', *Antiquity*, vol. 66, 1992, pp. 831–44.

Olejniczak, A.J., T.M. Smith, R.N.M. Feeney, R. Macchiarelli, A. Mazurier, L. Bondioli, A. Rosas, J. Fortea, M. de la Rasilla, A. Garcia-Tabernero, J. Radovčić, M.M. Skinner, M. Toussaint & J.J.

Hublin, 'Dental tissue proportion and enamel thickness in Neanderthal and modern human molars', *Journal of Human Evolution*, vol. 55, 2008, pp. 12–23.

Richards, M., K. Harvati, V. Grimes, C. Smith, T. Smith, J-J. Hublin, P. Karkanas & E. Panagopoulou, 'Strontium isotope evidence of Neanderthal mobility at the site of Lakoni, Greece using laser-ablation PIMMS', *Journal of Archaeological Science*, vol. 35, 2008, pp. 1251–6.

Stringer, C.B. & P. Andrews, 'Genetic and fossil evidence for the origin of modern humans', *Science*, vol. 239, 1988, pp. 1263–8.

Trinkaus, E., 'European early modern humans and the fate of the Neanderthals', PNAS, vol. 104, 2007, pp. 7367–72.

Watson, E., P. Forster, M. Richards, H-J. Bandelt, 'Mitochondrial footprints of human expansions in Africa', *American Journal of Human Genetics*, vol. 61, no. 3, 1997, pp. 691–704.

Yamei, H., R. Potts, Y. Baoyin, G. Zhengtang, A. Deino, W. Wei, J. Clark, X. Guangmao & H. Weiwen, 'Mid-Pleistocene Acheulean-like stone technology of the Bose Basin, South China', *Science*, vol. 287, 2000, pp. 1622–6.

19 'Vampire' project

Behar, D.M., R. Villems, H. Soodyall, J. Blue-Smith, L. Pereira, E. Metspalu, R. Scozzari, H. Makkan, S. Tzur, D. Comas, J. Bertranpetit, L. Quintana-Murci, C. Tyler-Smith, R.S. Wells, S. Rosset & The Genographic Consortium, 'The dawn of human matrilineal diversity', *American Journal of Human Genetics*, vol. 82, no. 5, April 24, 2008, pp. 1130–40.

Burke, H., C. Lovell-Jones & C. Smith, 'Beyond the looking glass: Some thoughts on sociopolitics and reflexivity in Australian archaeology', *Australian Archaeology*, no. 38, 1994, pp. 13–22.

Cann, H.M., C. de Toma, L. Cazes, M.-F. Legrand, V. Morel, L. Piouffre, J. Bodmer, W.F. Bodmer, B. Bonne-Tamir, A. Cambon-Thomsen, Zhu Chen, Jiayou Chu, C. Carcassi, L. Contu, R. Du, L. Excoffier, J.S. Friedlaender, H. Groot, D. Gurwitz, R.J. Herrera, Xiaoyi Huang, J. Kidd, K.K. Kidd, A. Langaney, A.A. Lin, S.Q. Mehdi, P. Parham, A. Piazza, M.P. Pistillo, Yaping Qian, Qunfang

Shu, Jiujin Xu, S. Zhu, J.L. Weber, H.T. Greely, M.W. Feldman, G. Thomas, J. Dausset & L.L. Cavalli-Sforza, 'A human genome diversity cell line panel', *Science*, vol. 296, no. 5566, Apr. 2002, pp. 261–2.

Li, J.Z., D.M. Absher, H. Tang, A.M. Southwick, A.M. Casto, S.Ramachandran, H.M. Cann, G.S. Barsh, M. Feldman, L.L. Cavalli-Sforza & R.M. Myers, 'Worldwide human relationships inferred from genome-wide patterns of variation', *Science*, vol. 319, no. 5866, 2008, pp. 1100–4.

Orlando, L., P. Darlu, M. Toussaint, D. Bonjean, M. Otte & C. Hänni, 'Revisiting Neanderthal diversity with a 100,000 year old mtDNA sequence', *Current Biology*, vol. 16, no. 11, 2006, pp. R400–2.

Smith, L., 'What is this thing called postprocessual archaeology ... and is it relevant for Australian archaeology?', *Australian Archaeology*, no. 40, 1995, pp. 28–31.

Stone, L., P.F. Lurquin & L.L. Cavalli-Sforza, *Genes, Culture and Human Evolution: A synthesis*, Blackwell Publishing, Oxford, 2007.

HUGO, 'The Human Genome Diversity (HGD) Project Summary Document', 1994.

Wells, S., *Deep Ancestry: Inside the Genographic Project*, National Geographic, Washington, DC, 2006, p. 247.

Wheeler, D.A., M. Srinivasan, M. Egholm, Y. Shen, L. Chen, A. McGuire, Wen He, Yi-Ju Chen, V. Makhijani, G.T. Roth, X. Gomes, K. Tartaro, F. Niazi, C.L. Turcotte, G.P. Irzyk, J.R. Lupski, C. Chinault, Xing-zhi Song, Yue Liu, Ye Yuan, L. Nazareth, Xiang Qin, D.M. Muzny, M. Margulies, G.M. Weinstock, R.A. Gibbs & J.M. Rothberg, 'The complete genome of an individual by massively parallel DNA sequencing', *Nature*, vol. 452, April 2008, pp. 872–6.

20 Back to country

Budinich, M., E. Montagnari & C. Tuniz, International Workshop on Science for Cultural Heritage, Trieste, Italy, October 23–27, 2006, Proceedings, World Scientific, Hong Kong (in press).

Chalmers, N. 'Statement of dissent from Neil Chalmers', in 'Report of the working group on human remains', 2003 <www.culture.gov.uk/images/publications/wghr_reportfeb07.pdf>.

Mulvaney, D.J., 'Past regained, future lost: The Kow Swamp Pleistocene burials', *Antiquity*, vol. 65, no. 246, 1991, pp. 12–21.

Smith, L., 'The repatriation of human remains—problem or opportunity', *Antiquity*, vol. 78, pp. 404–13.

Tafforeau, P. & T.M. Smith, 'Nondestructive imaging of hominoid dental microstructure using phase contrast X-ray synchrotron microtomography', *Journal of Human Evolution*, vol. 54, 2008, pp. 272–8.

Acknowledgements

Writing this book involved interviews with some of the world's top researchers, many of whom welcomed us into their laboratories and onto sites. However, space did not permit coverage of all scholars who have made big contributions to palaeostudies.

Some of the results and hypotheses reported here will be refined—perhaps even abandoned—as more evidence comes to hand. The opinions in this book are ours, and are not necessarily shared by those we acknowledge here.

We thank Jeremy Austin, Alex Baynes, Jim Bowler, Peter Brown, John Chappell, Ron Clarke, Alan Cooper, Matt Cupper, Francesco d'Errico, James Dixon, Louise Dunn, Graham Farquhar, Keith Fifield, Tim Flannery, Peter Forster, Michael Gagan, Russ Graham, Colin Groves, Rainer Grün, Simon Haberle, Hal Hatch, Harvey Johnston, Rhys Jones, Peter Kershaw, John Long, Simon Longstaff, Luigi Luca Cavalli-Sforza, Ernest Lundelius, John Magee, Paul Martin, Duncan Merrilees, Gifford Miller, John Mitchell, Junette Mitchell, Leanne Mitchell, Mike Morwood, John Mulvaney, Jon Olley, Tim Partridge, Svante Pääbo, Gavin Prideaux, Tom Rich, Richard Roberts, Mike Smith, David Steadman, Chris Stringer, Grant Sutherland, Phillip Tobias, Chris Turney, Steve Webb, Elizabeth Weiss, Rod Wells, Spencer Wells, Michael Westaway, Martin Williams and Milford Wolpoff.

Index

Page numbers in *italics* refer to figures.